P9-BIM-859

THE LOCUST

Also by Robert Barrass

Biology: Food and People

the economic importance of biology

THE LOCUST

A Guide for Laboratory Practical Work

ROBERT BARRASS

B.Sc., Ph.D., F.I.Biol
Sunderland Polytechnic

BARRY SHURLOCK

BARRY SHURLOCK
& Co. (Publishers) Ltd
174 *Stockbridge Road*
Winchester
Hants SO22 6RW

First published 1964
Reprinted 1968
Second edition 1974

ISBN 0 903330 11 3

To

S. H. J.

Typeset and printed
in Great Britain by
REDWOOD BURN LIMITED
Trowbridge & Esher

Contents

Preface

This guide introduces students to the features of arthropod organization and places considerable emphasis on work with living animals. In its approach it differs from many other laboratory guides since the student is encouraged to make observations and carry out experiments, to make notes and drawings, and to record his own findings. So that the student will always be aware of the living insect, experiments are included in several parts of the text. These experiments are taken from the published work of research scientists. A reference to the original paper is given with each experiment, for those who may require further information.

The locust is selected for study because it is a large insect which can be reared easily and cheaply throughout the year. Its large size makes it well suited for a study of the anatomical characteristics of arthropods and insects. Living locusts are a source of great interest in the laboratory; the life-cycle is short and all stages can be observed.

The locust is preferable to the cockroach, which is still sometimes used for dissection. The nocturnal habits, long life-cycle, and dorso-ventral flattening of the cockroach, make it less suitable for students looking at insects for the first time. Cockroaches also infest kitchens and food stores, and many teachers are reluctant to keep them alive for this reason.

This book is intended primarily for students in the last two years at school (up to 'S' level G.C.E.) or in the first year of a College or University course.

R. B.

Kingston-upon-Hull
March, 1964

Preface to Second Edition

Twelve years ago living locusts were not available commercially in Britain, and in preparing this guide I was able to interest only one firm in locust breeding as a commercial proposition. Now this firm sells more locusts than any other laboratory animal, and most biological supply firms keep large stocks. Locusts are also reared in small numbers in many schools and colleges. In the ten years since the publication of the First Edition, the usefulness of the locust as a laboratory animal has been recognised by many teachers, particularly those interested in teaching biology by observation and investigation.

In countries where locusts are not readily available, this guide may still be used. The anatomy of the locust is similar to that of other grasshoppers, and many of the investigations suggested here could be undertaken with any locust or grasshopper.

I should like to thank those readers who made suggestions for improving the First Edition. Additional practical exercises have now been included, together with further suggestions for project work.

R. B.

Sunderland
June, 1974

Acknowledgments

For their help in reading and discussing the manuscript and for many valuable suggestions, I wish to thank the following: Sir Boris P. Uvarov, Dr P. T. Haskell, Mr Philip Hunter-Jones, Dr Maud J. Norris, Dr D. F. Woodrow, all of the Anti-Locust Research Centre; Dr J. D. Carthy, Queen Mary College, University of London; Mr C. G. Gardener, Birmingham College of Advanced Technology; Mrs H. B. Miles, Hull Municipal Training College; Mr T. C. Dunn, Chester-le-Street Grammar School; Miss E. J. Harris, St. Mary's Grammar School, Hull; and Mr B. Tear, South Park High School, Lincoln.

I also wish to thank my colleagues at the Kingston-upon-Hull College of Technology, Mr S. Cobb, Mrs G. S. de Boer and Dr T. A. G. Wells, for their help and advice; Mr C. Bullock, for trying out many of the techniques; Mr O. C. Gay, for allowing me to make use of photographic facilities in the Department of Physics; and Mr D. A. Moore, Dr D. C. White, Mr E. Worsley and Mr B. A. Leadbeater, for their advice and help with the photography.

My wife, Ann, has helped in preparing this book and given constant encouragement.

I am also indebted to the staff of my publishers, for their help in preparing this book for the press.

Introduction

Students working with locusts will want to know something of their habits and distribution and, because this information is not readily available in textbooks, an introduction is included here.

The Locust Problem

There are some 5,000 different species of grasshopper. A few of these, which sometimes occur in tremendous numbers, are called locusts. These few species are, usually, no more closely related to each other than to other species of grasshopper. When locusts live in such crowded conditions they differ in both appearance and behaviour from solitary individuals of the same species.

A locust eats approximately its own weight in vegetation every day and a swarm of several million locusts consumes many tons of food in a day. In this way, man's crops are sometimes devoured over vast areas and widespread famine may result. Famine is all the more likely when, in times of food shortage, the people of these tropical countries are already living on a minimum subsistence diet. Locust plagues are an ever-present danger to agriculture, and therefore to the lives of men, in many tropical and subtropical countries. There is nothing new in this.

> 'For they covered the face of the whole earth, so that the land was darkened; and they did eat every herb of the land . . . '
>
> Exodus 10; 15.

More recently the economic losses due to locust plagues have been estimated as being in the region of £30 million a year (Uvarov, 1951).

Distribution (Figure 1)

Two species of locust which cause perhaps the greatest financial loss are the desert locust (*Schistocerca gregaria*) and the

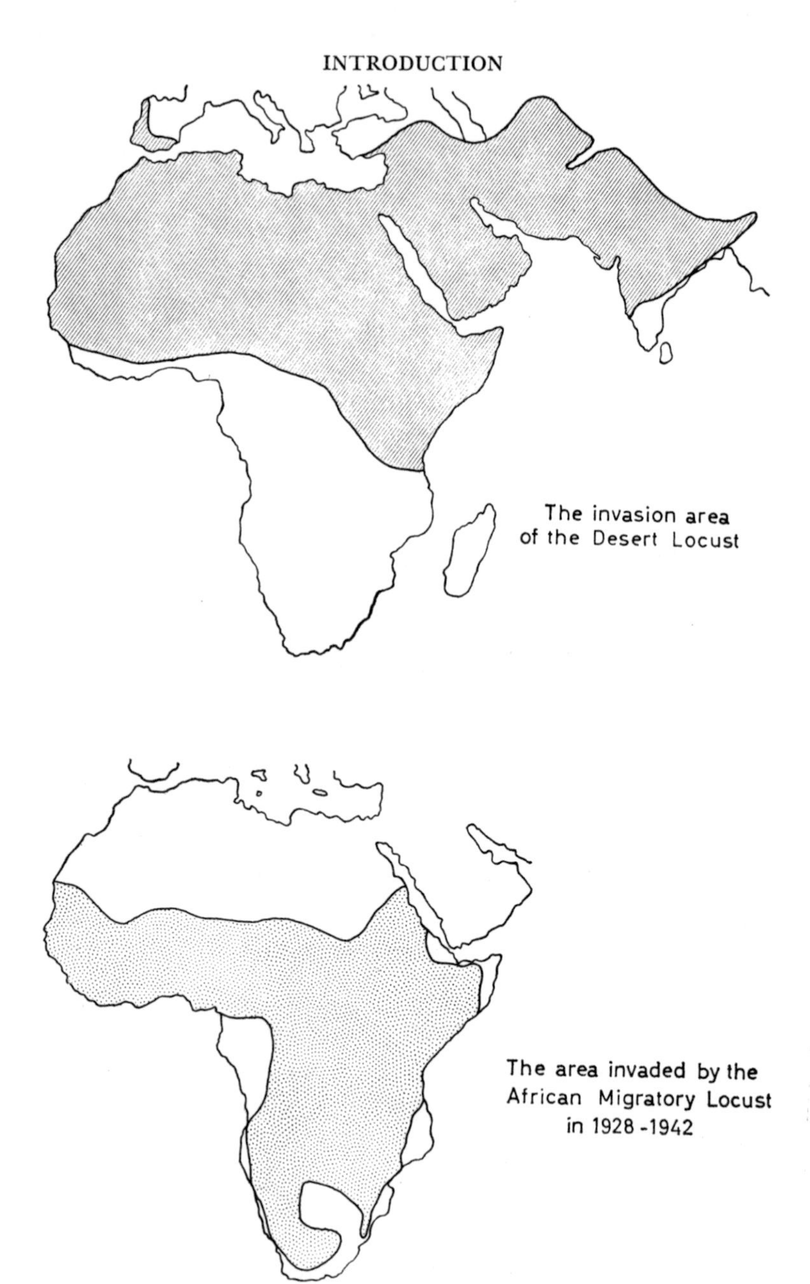

Figure 1. Geographical distribution of the Desert Locust (above) *and the African Migratory Locust* (below). (Based on information supplied by The Centre for Overseas Pest Research)

migratory locust (*Locusta migratoria*). The desert locust occurs in the northern half of Africa and as far east as northern India. The different sub-species of the migratory locust have a range extending from west and south Europe and most of the African continent, through central and southern Asia to China and northern Australia. Over much of this area the migratory locust looks and behaves like any solitary grasshopper but in some places, within its geographical range, swarms may arise which cause extensive devastation. To understand how this comes about it is necessary to know something of the life-history.

Life-history

The female locust, using the hard valves of its egg-laying organ, digs a hole about 10 cm deep in the sand. Eggs are laid in the lower two thirds of this hole, and mixed with a frothy secretion which hardens to form a tubular 'egg-pod'. The upper third of the hole is filled with the frothy secretion only. Embryonic development takes about two weeks at 28–33°C. The nymphs emerging from the eggs are vermiform (worm-like) larvae because they wriggle up through the egg-pod to the surface of the sand.

The skin or cuticle of an insect is also its skeleton; it not only covers and protects the body but also provides a surface to which muscles are attached. Because this cuticle is inflexible when hardened, only a very limited amount of growth is possible, and insects periodically cast the cuticle, or moult, in the course of their nymphal development. After each moult the new cuticle is soft and the locust increases in size before the cuticle hardens.

The vermiform larva moults on reaching the surface of the sand and it is then called a hopper because of its method of progression. The hopper sheds its exoskeleton five more times, over a period of several weeks, before it reaches the adult, winged stage. After each moult (ecdysis) the nymphal form is called an instar and the period of time between moults is called a stage or stadium. In the first three hopper instars the developing wings are small ventrally directed lobes; in the fourth and fifth they point posteriorly and the fore-wings are covered by the hind wings (*Figure 7*).

Reproductive Potential

An egg-pod contains between 70 and 100 eggs and a female locust may lay several pods of eggs in the course of its life. It is possible, therefore, for many hundreds of nymphs to emerge from the eggs laid by a single female. Usually many of these offspring fail to survive to maturity and the number of locusts in an area does not change very much from one generation to the next. Sometimes, however, large numbers of offspring do reach maturity and they, in turn, lay many eggs. When this happens a tremendous increase in population size takes place and a plague may develop.

'Outbreak' Areas

Plagues of *Locusta* originate in comparatively small areas within the total range of distribution of the species. These areas, in which environmental conditions favour the rapid increase in numbers of locusts, are called 'outbreak' areas. An essential feature of an 'outbreak' area (Uvarov, 1951) is that the environment is subject to sharp fluctuations in weather and vegetation. Rainfall or flooding result in a sudden luxuriant growth of vegetation and a rapid increase in the numbers of locusts. When dry conditions follow, patches of vegetation become smaller and the locusts are crowded together in these more restricted habitats.

Gregarious and Solitary Phases

Crowding of the hoppers affects the colour, structure and behaviour of both the hoppers and the ensuing adults. In Nature, locusts are much more active when crowded together than when they are alone and they reach maturity more quickly. Locusts in a crowd move about as gregarious marching bands of hoppers or as dense flying swarms of adults. These are the locust plagues which devour vegetation and are a problem to man. A swarm of 40 000 million locusts, covering 1200 km^2, may consume 80 000 tonnes of food in one day.

When locusts are not crowded (that is, they are in the solitary phase) they look different in form and colour (*Figure 7*) from individuals in the gregarious phase (*see* Experiments 6 and 7).

The difference between the solitary and gregarious phases of the same species is so marked that they were once considered different species and given different names. It must be emphasized, however, that we do not yet have a complete understanding of the change from one phase to the other.

Control

When swarms of locusts leave the 'outbreak' areas the locust plague has started and it is not easy to control. Locusts may fly hundreds of miles, moving with the prevailing wind, and on settling they feed on natural vegetation or agricultural crops. They lay eggs which may give rise to the solitary or the gregarious phase, according to the number of hoppers reaching maturity and the environmental conditions. In this way new breeding areas may be formed far away from those in which the plague originated.

The earliest control measures were attempts to kill locusts when they had invaded crop areas. Beating the insects with sticks was an early but ineffective means of control. The numbers are too great for this. Knowledge of the life-history of the locust and an understanding of the nature of outbreak areas has led to more effective measures. Regular patrols of the 'outbreak' areas by field officers give information about the number of locusts in any area. In this way the breeding population is watched carefully.

When, in any area, population increases are noted, control measures are brought into operation before the numbers reach uncontrollable limits. The aim is to prevent an outbreak. Usually poison baits or insecticidal sprays and dusts are used in an attack on the hoppers. If the locusts fly, they can be attacked in the air. Aeroplanes flying through the swarm spray insecticides and by repeated efforts, gradually reduce the size of the swarm.

Despite man's careful watch, locust outbreaks still occur. The long term solution to the locust problem may involve the use of ecological methods to alter environmental conditions in the 'outbreak' areas and so prevent swarming.

Locusts do not respect international boundaries and a most hopeful aspect of locust control is that international co-operation is now well established. Many countries co-operate in sending information, gathered by their field officers, to the Centre for Overseas Pest Research, London. This information, re-issued in the form of monthly situation reports, is used in international control. Co-ordination of the efforts of different nations is arranged through the Food and Agriculture Organization of the United Nations in Rome.

Rearing Locusts

THE MIGRATORY LOCUST

Locusta migratoria migratorioides R. & F., the African Migratory Locust, may be reared easily and cheaply throughout the year. A cage (*Figure 2*) large enough for up to 300 locusts, can be constructed fairly cheaply from hardboard bolted to a frame of aluminium angle strips. Hardboard is used for the sides, back, floor and roof. A trap door is cut into the roof to permit the introduction of food, the removal of waste or the handling of locusts. The false floor is of perforated zinc; this helps in the ventilation of the cage and permits the easy placing of the jars of sand in which eggs are laid. The front of the cage, extending from the false floor to the roof, is a glass observation window. A similar cage can be made very cheaply by converting a half tea-chest.

Heat and Light

A light bulb should be placed in the cage roof between the trap door and one of the corners. This bulb should give a day-time temperature of 28–34°C in the various parts of the cage. The wattage of this light bulb will therefore depend upon the temperature of the laboratory. In cold weather a bulb below the false floor may also be necessary if the locusts are to continue to breed. This bulb should be enclosed so that it provides heat but not light. Locusts are most healthy in alternating day and night conditions so the cage light should be switched off each evening.

Moulting

Long branching twigs or plastic netting should be placed across the cage from corner to corner, so that nymphs can hang head downwards when they are moulting.

Feeding

Fresh (but not wet) grass provides an adequate diet. Over the week-end, dry wheat bran, dried grass and carrot peelings are useful additional foods. When only dry food is available, invert a beaker of water on a cotton wool pad in a petri dish. The faeces should be dry; if not this indicates that the diet contains too much water.

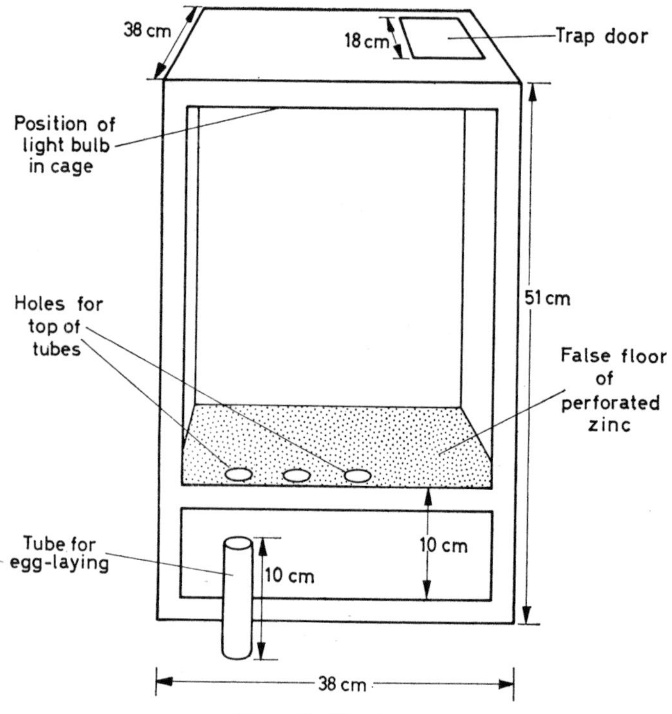

Figure 2. Rearing cage

Cleaning

Old grass, faeces and dead locusts should be removed each day. A small vacuum cleaner may be used. This reduces the time needed and also makes less dust. At intervals, the cage should be scrubbed and sterilized with 0·5 g chlorocresol in 100 cm^3 water.

Egg-laying

Mature females will lay eggs in a 1 lb. jam jar (or in a metal container, *see Figure 2*) full of sand. Mix five parts of clean, dry, sterilized sand with one part of water. Place the jar under the false floor so that the sand surface is level with the cage floor. Incubate the eggs at 28–33°C and keep the jar covered with a loose-fitting lid. The eggs will hatch in about two weeks. If the sand is too fine or if too much water is used, the sand will be waterlogged and the eggs will not live.

THE DESERT LOCUST

Schistocerca gregaria Forsk., the desert locust. Always provide dry wheat bran. Place the incubated eggs in the rearing cage just before they are expected to hatch. For greater detail on locust rearing techniques *see* Hunter-Jones (1961).

Note: Some people develop an allergy when keeping locusts, and this may be apparent as a cutaneous or nasal reaction. Technicians handling locusts continually may wish to wear protective clothing. Write for further information to the Centre for Overseas Pest Research, College House, Wrights Lane, London, S.W.7.

Classification

ARTHROPOD CHARACTERISTICS

Locusts belong to the Phylum ARTHROPODA and to the Class INSECTA. Examine the imago (adult); the following external features are characteristic of all arthropods (*Figure 3*):

(1) The paired jointed appendages (arthro=joint; poda=feet).
(2) The cuticle or exoskeleton (exo=on the outside) which covers the body and functions as both a skin and a skeleton.
(3) The flexible and elastic articular membranes which separate the hard plates (sclerites) of the exoskeleton.
(4) The division of the body into segments.
(5) The bilateral symmetry.

(*See* page 40 for internal features.)

INSECT CHARACTERISTICS

Note also the following insect characteristics (*Figure 3*):

(1) The head with a pair of compound eyes and between them three simple eyes or ocelli, a single pair of antennae and mouthparts comprising an upper lip or labrum, a pair of mandibles, a pair of maxillae and a lower lip or labium (*see Figures 10* and *11*).
(2) The thorax of three segments designated pro-, meso- and metathoracic segments (*see Figure 12*), each with a pair of ventro-lateral legs; for this reason the insects are sometimes called HEXAPODA (=six feet). Wings are only present in the imago; they are dorso-lateral expansions of the meso- and metathorax.
(3) The abdomen of eleven segments. The genital opening is near the anus. The only appendages are the external genitalia and terminal cerci (*see Figure 8*). On each of the first eight segments there is an opening (spiracle) into the tracheal system.

(*See* page 40 for internal features.)

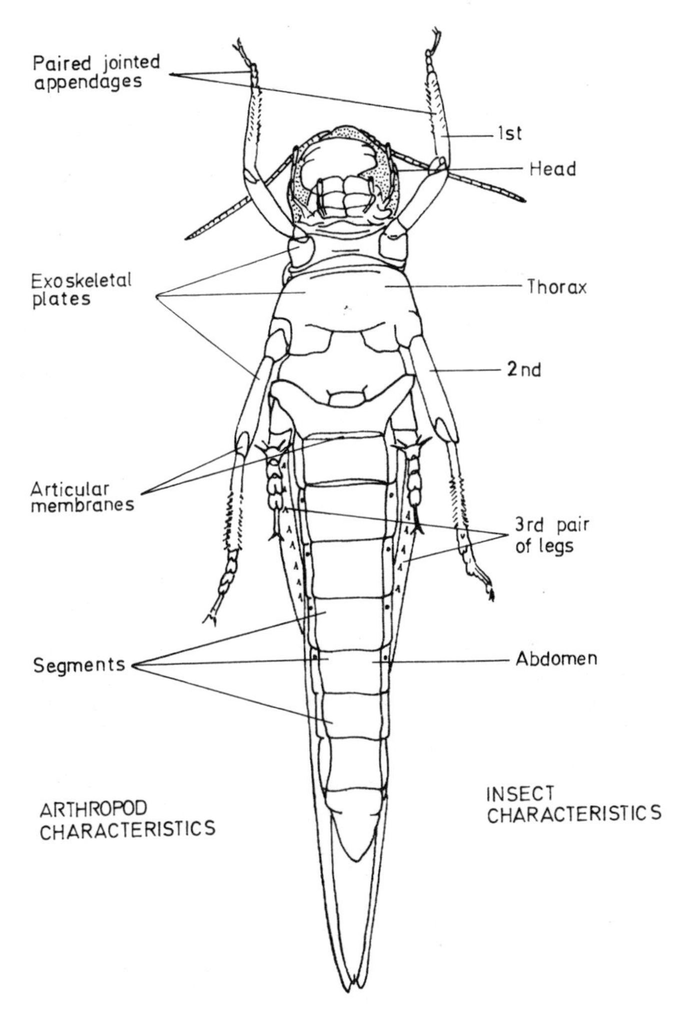

Figure 3. Ventral view of a male Locusta *labelled to show features characteristic of all arthropods (note also that the left side of the body is essentially similar to the right side: the animal is bilaterally symmetrical), and also to show features characteristic of all insects.* (Magnification $\times 2$)

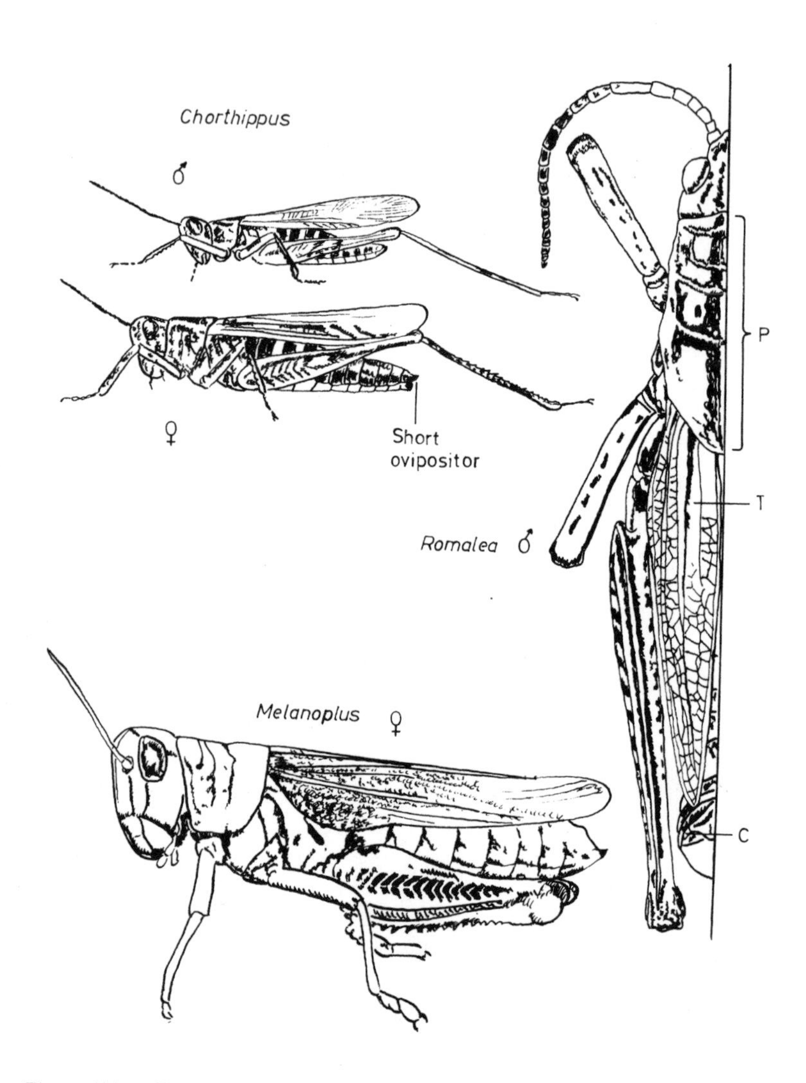

Figure 4(a). Short-horned grasshoppers (Superfamily: Acridoidea) are included in the Order Orthoptera. P = *large prothorax;* T = *rather hard fore-wing or tegmina;* C = *short cercus.* (Magnification ×2)

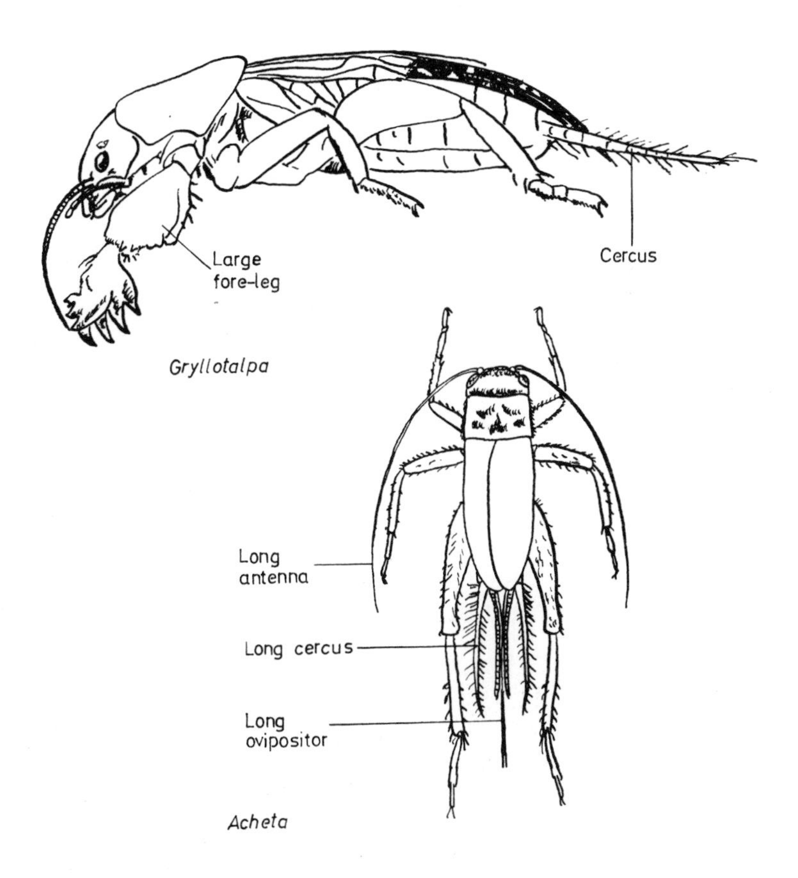

Figure 4(b). Crickets (Superfamily: Grylloidea) and mole crickets (Superfamily: Gryllotalpoidea) are also included in the Order Orthoptera. (Magnification × 2)
Note that the stick insects and leaf insects (Order: Phasmida) and the cockroaches and mantids (Order: Dictyoptera), which are believed to be descended from the same ancestral stock as the grasshoppers and crickets, are not now included in the Orthoptera

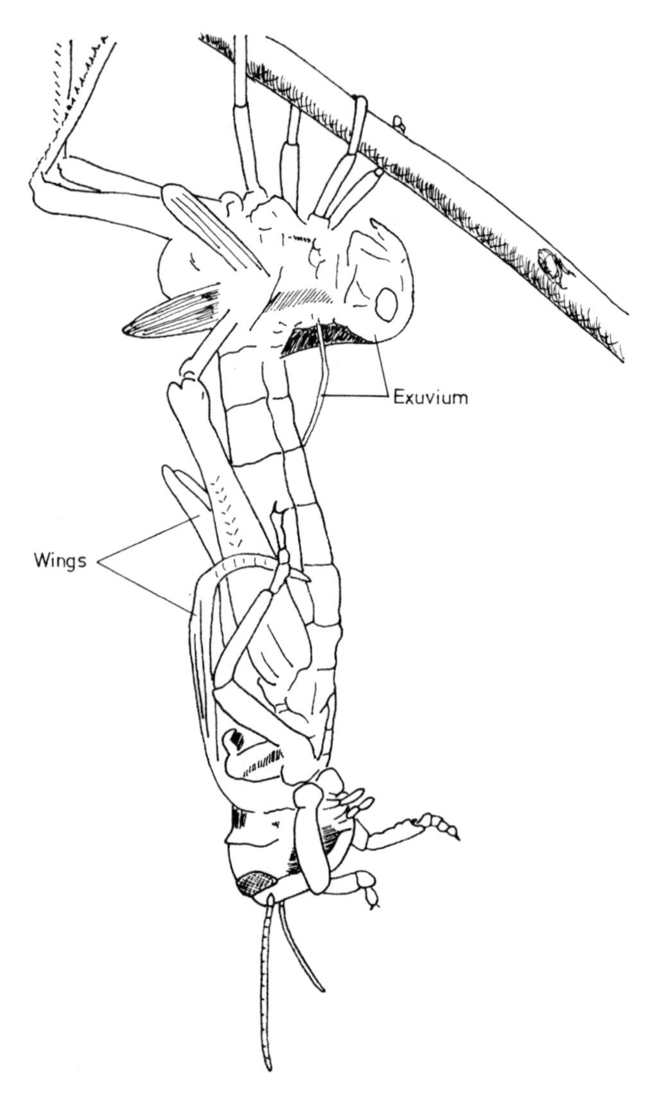

Figure 5. Ecdysis in Locusta. *Imago hanging head downwards, and still attached to the exuvium of the 5th hopper stage.* (Magnification $\times 2$)

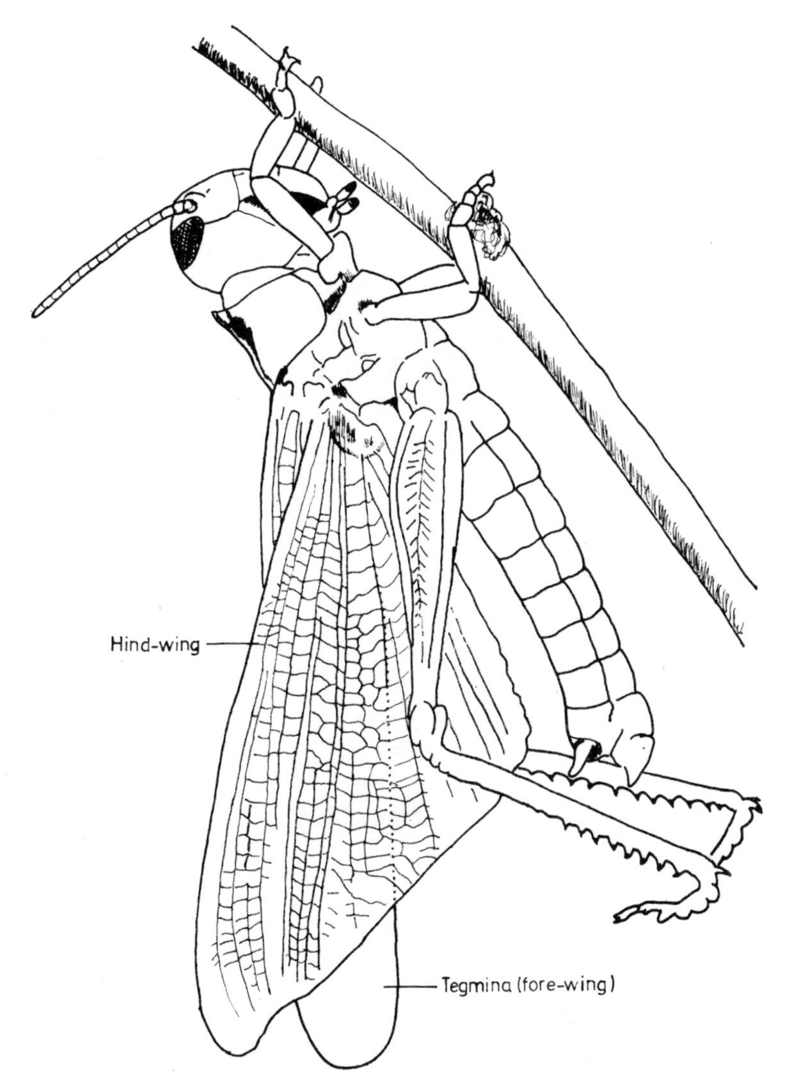

Figure 6. Locusta: *newly fledged imago rests with head up as its wings expand.* (Magnification × 2)

Based on colour transparencies from Harris Biological Supplies (See *p. 63*)

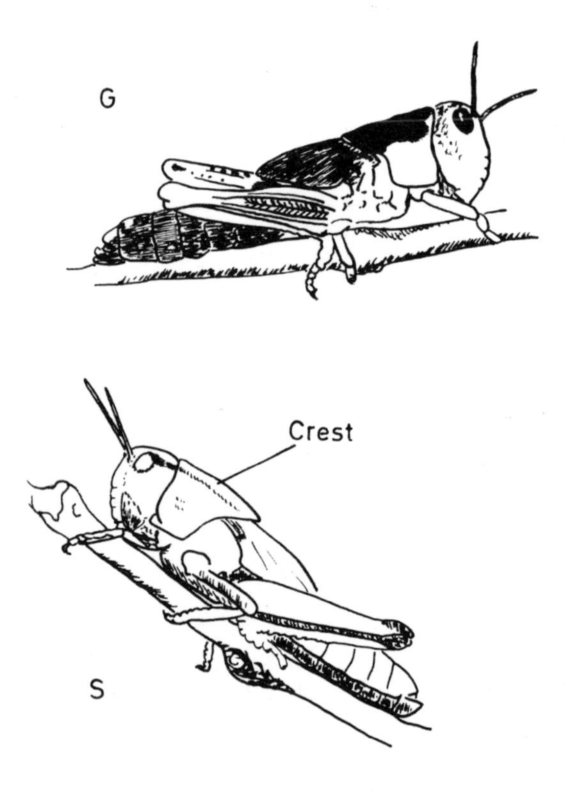

Figure 7. Locusta *5th hopper stage:* S = *solitary phase* (*green and fawn*), G = *gregarious phase* (*black and orange*). *Intermediate forms, between these two extremes, also occur* (*phase* transciens): *an example of continuous polymorphism. From Barrass, R.* (*1974*). Biology: Food and People. London; English Universities Press.

INSECT LIFE-CYCLES

The Orders of winged insects are grouped into two divisions according to their type of life-cycle.

(1) EXOPTERYGOTA (developing wings apparent externally). This division includes the locust and also stick insects, dragonflies, earwigs, cockroaches, termites, lice and bugs. These insects hatch from the egg as nymphs which, in body form, are in many respects similar to the imago. The type of food ingested is similar at all stages. There is no pupal stage.

(2) ENDOPTERYGOTA (developing wings not apparent externally). This division includes the butterflies, flies, fleas, bees and beetles. The body form of the larva is very different from that of the imago and they consume different kinds of food. Transformation from the larval to the imaginal body form (metamorphosis) takes place in a pupal stage.

GENUS *LOCUSTA*

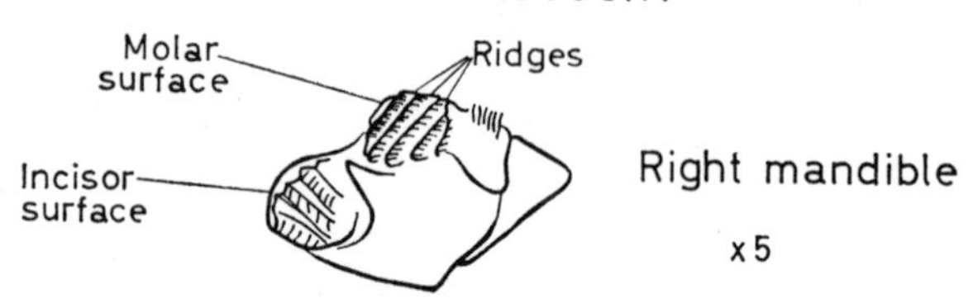

1. No peg-like process between coxae of fore-limbs
2. Thoracic sterna covered in very fine 'hairs'

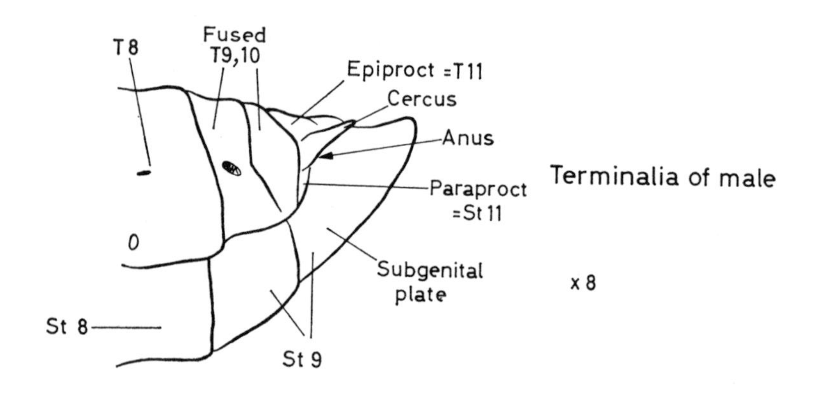

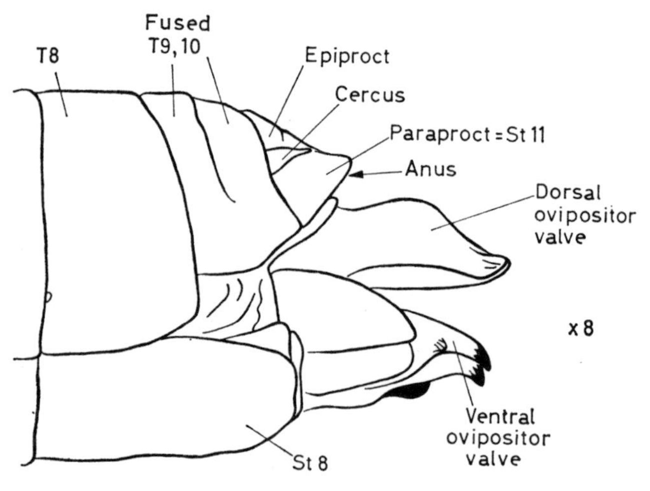

Terminalia of female

Figure 8(a). Some features of the genus Locusta

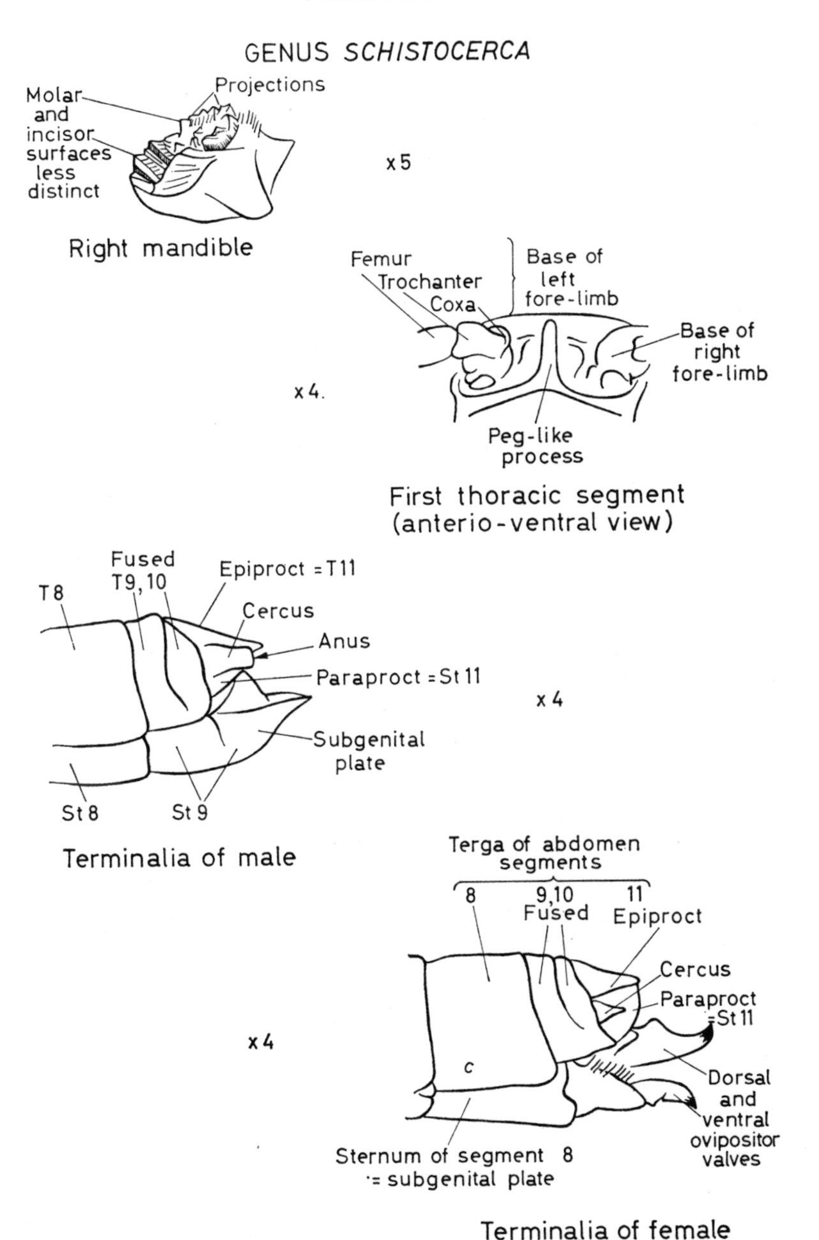

Figure 8(b). Some features of the genus Schistocerca

SUMMARY OF CLASSIFICATION

Phylum: Arthropoda—with an exoskeleton and jointed appendages.

Class: Insecta—body divided into head, thorax and abdomen. Six legs.

Sub-class: Pterygota—having wings.

Division: Exopterygota—wings develop externally, no pupal stage, young called nymphs (*Figures 5* and *7*).

Order: Orthoptera—(orthos = straight; pteron = a wing), with a large prothorax; hind legs used in jumping; fore-wings rather hard and called tegmina; with cerci (*Figure 4*).

Superfamily: Acridoidea—short horned grasshoppers.

Genus: There are several genera of locusts. Two examples are *Locusta* and *Schistocerca*; they differ in the following respects.

Locusta: With a distinct difference between the incisor and molar surfaces of the mandibles; with a 'hairy' underside to the thorax; with no peg-like process between the bases of the fore-limbs and with rather rounded external genitalia in both sexes.

Schistocerca: With no distinct difference between the incisor and molar surfaces of the mandibles; underside of thorax without 'hairs'; with a peg-like process between the bases of the fore-limbs and with rather pointed external genitalia.

These differences are illustrated in *Figure 8*.

Note that the features used in the above classification are external ones. Different insects also differ in their internal anatomy but these differences are not usually used in classification since it is often necessary to identify an insect without killing it. Additional reasons for basing a classification on external anatomy are: (1) dissection is time-consuming; and (2) the hard external parts are most easily preserved. Methods for the collection, preservation and study of insects are given by Oldroyd (1970).

SEX DETERMINATION

View the terminalia from below (*Figure 9*).

Female: in the first hopper instar two pointed plates (the upper ovipositor valves) can be distinguished. In the second,

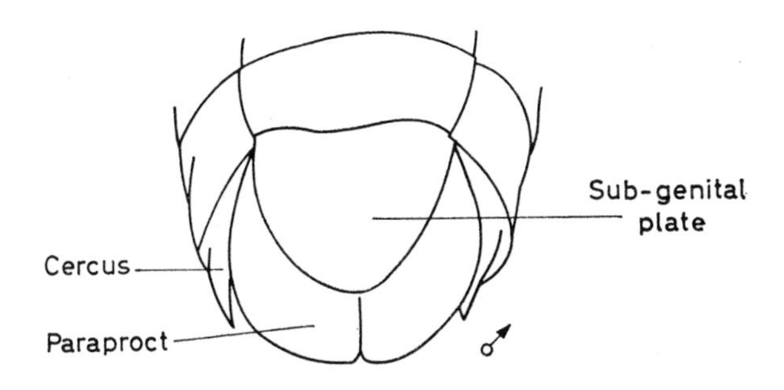

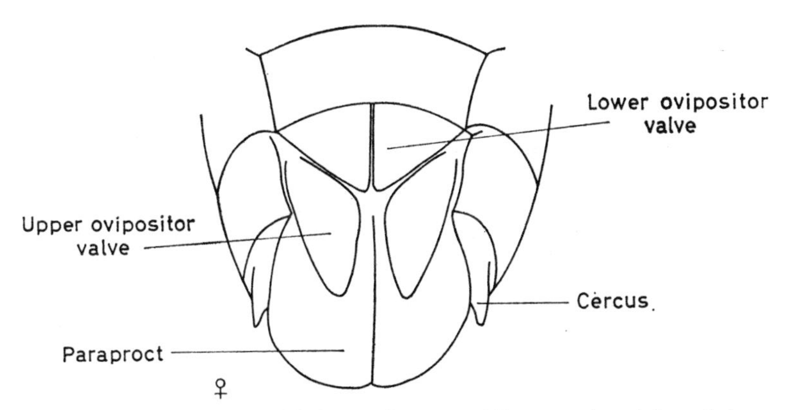

Figure 9. Terminalia of third stage hoppers of Locusta *viewed from below: male* (above) *and female* (below). (Magnification ×25)

there are two more anteriorly placed plates (the lower ovipositor valves). In later instars these valves are more distinct.

Male: the subgenital plate is bilobed in the first hopper instar, less so in the second and is not bilobed in later instars (Dirsh, 1950). The genitalia of the imago are shown in *Figure 8*.

Experiments

In any experiment always use locusts which have all been treated in the same way and are at the same stage of the life-history. As far as possible, keep conditions constant throughout an experiment. Even if this is done, there will still be some variation in behaviour. If time is available repeat each experiment several times and average the results (or those of the whole class). Wherever possible express the findings numerically but do not lose sight of the behaviour of individuals.

EXPERIMENTS ON DEVELOPMENT

Experiment 1. Duration of Development

If an incubator is available find the effect of temperature on the rate of development of the eggs. Do not allow the sand in the jars to become too dry (cover the jars with a loose-fitting lid).

Experiment 2. Weight of Developing Eggs

Weigh ten eggs on each day of incubation. Plot the mean weight for each day on a graph. Between weighings keep the eggs in a mixture of five parts of dry sand to one part of water. Do the eggs take up or lose water? (*See* Hunter-Jones, 1964.)

EXPERIMENTS ON GROWTH AND ECOLOGY

Experiment 3. Feeding and Growth of Hoppers

After a vermiform larva has moulted, weigh it, measure the length of the head and body, and then place it in a 2 lb. jam jar with food. Repeat the weighing and measurements every 24 h until it is adult (i.e. in about 1 month). Also dry and then weigh the faeces produced each day. Tabulate the results. Does the hopper consume similar amounts of food (as evidenced by the production of faeces) on each day of an instar? Do head length and body length increase evenly from day to day? Divide head length into body length: is the ratio approximately the same for the different instars?

Experiment 4. Moulting

Observe fifth stage hoppers in the rearing cage. Just before ecdysis they hang head downwards from twigs. Observe ecdysis (*Figure 5*) and describe what you see. After ecdysis the insect hangs head-up as its wings expand (*Figure 6*).

Experiment 5. Increase in Numbers

Devise a simple experiment to determine how many egg-pods a solitary female produces in its life. Multiply this figure by the number of eggs in a pod. This result gives an indication of the potential increase in numbers from one generation to the next.

Experiment 6. Colour of Solitary Phase

Rear ten first stage hoppers in separate 2 lb. jam jars. Put a branching twig in each jar and cover the bottom with dry sand. Provide a little fresh (but not wet) grass each day. Repeat with ten more hoppers but provide plenty of fresh grass, standing in a tube of water. Compare the colours of all stages.

Experiment 7. Colour of Gregarious Phase

Rear ten hoppers in a 2 lb. jam jar. Compare the colour of the nymphal stages and the newly moulted adults with those of the solitary phase.

Preserve the insects from experiments 6 and 7 and, when you have studied the external anatomy of the thorax (pp. 28–31), compare the shape of the pronotum of crowded and solitary forms; and with each one divide the length of the femur of the hind limb into the length of the fore-wing.

Experiment 8. Population Size Estimate

Catch 20 locusts from the breeding cage and mark them on the under surface of the abdomen with a spot of quick drying paint. Release them into the cage. Half an hour later, remove 25 locusts, examine them and observe how many of them are marked (say 5). Since:

$$\frac{\text{marked locusts in sample}}{\text{total locusts in sample}} = \frac{\text{total marked locusts}}{\text{total locusts in cage}}$$

then the number of locusts in the cage may be estimated as:

$$\frac{5}{25} = \frac{20}{n}; n = \frac{25}{5} \times 20 = 100 \qquad \text{[Jackson (1933)]}$$

External Features: Form and Function

Use a good hand lens or, if possible, a binocular dissecting microscope.

THE HEAD

Draw the head to show the exoskeletal plates (*Figure 10*), two many-jointed antennae, two compound eyes (each with many facets), three ocelli (each with a single lens), and the mouthparts. Hold down the locust, on its dorsal surface, by pinning through the legs onto the wax of a dissecting dish; hold the labrum forwards and note the hairs on its inner surface; examine the mandibles, note that they are asymmetrical and that each one has both a cutting (incisor) and a grinding (molar) surface (*Figure 8*); note the parts of the maxillae (*Figure 11b*) and labium (*Figure 11a*). Hold back the labium and note the centrally placed, tongue-like, hypopharynx (*Figure 11b*).

The difference between the incisor and molar surfaces of the mandibles is more marked in *Locusta* (which feeds on grasses) than in *Schistocerca* (which feeds mostly on broad leaved plants). The head is attached to the first thoracic segment (prothorax) by a cervical membrane (cervix = neck). An occipital condyle on each side of the head capsule articulates with a first cervical sclerite, and a second cervical sclerite articulates with the episternum of the prothorax.

Place the head in a test tube in 10 per cent sodium hydroxide to remove the soft internal parts (24 h in cold sodium hydroxide or less than 30 min if heated in a hot water bath). Pin the head through the foramen on to the wax of a dissecting tray and then draw the head in posterior view (*Figure 11a*). Raise the labium with forceps and note the hypopharynx (*Figure 11b*). Great care is needed in removing the labium or the hypopharynx will be torn away at the same time. Hold the labium with forceps and remove it with a mounted cutting needle; draw (*Figure 11b*).

To remove the maxillae, hold each at its base and carefully cut it away. Use forceps and a cutting needle to remove the mandibles before attempting to remove the hypopharynx. Make separate permanent mounts (methods are given below) of the labium, maxillae, hypopharynx, an antenna and part of the

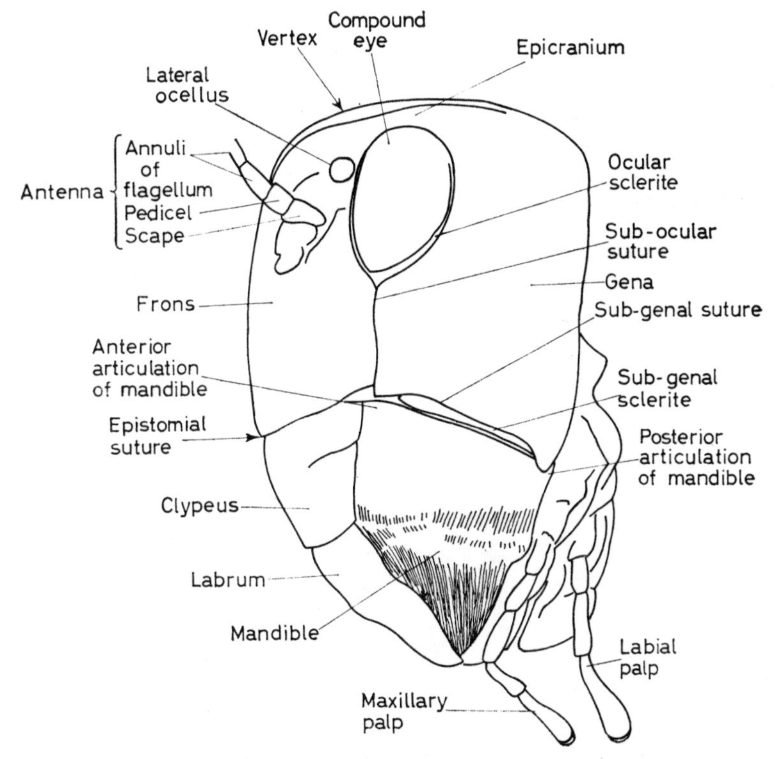

Figure 10. Head of Locusta; *view from left side.* (Magnification ×9)

cuticle (cornea) covering the compound eye. Look through the ventral hole in the head capsule (mouth parts removed) at the internal cuticular ridges. Cuticular inpushings (apodemes) form the anterior and posterior tentorial arms and converge on the central body of the tentorium (internal skeleton of the head).

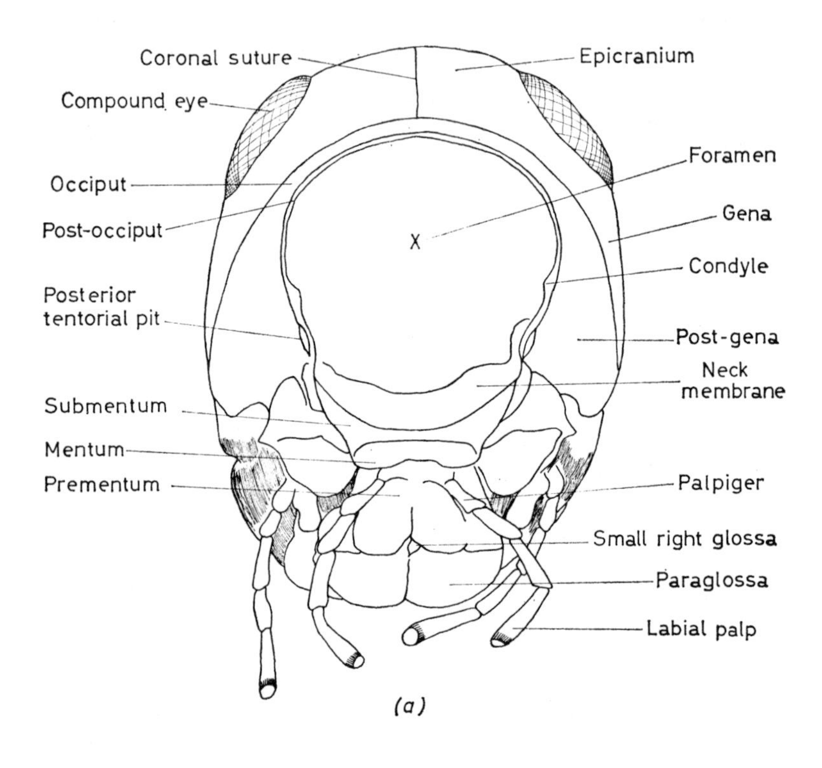

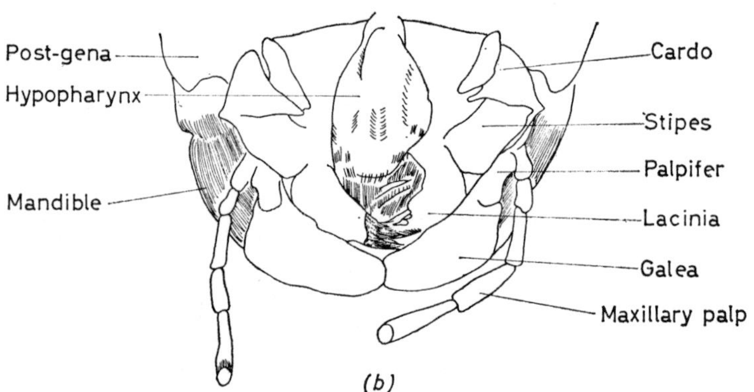

Figure 11. Head of Locusta: *(a) posterior view showing parts of labium; (b) mouthparts after removal of labium.* (Magnification × 9)

Preparation of Permanent Slides of the Mouthparts

(1) Wash in 70 per cent then in 90 per cent alcohol;
(2) transfer to 95 per cent alcohol (5 min),
(3) to 100 per cent alcohol (at least 5 min),
(4) to fresh 100 per cent alcohol (5 min),
(5) to xylene (at least 10 min);
(6) mount on separate slides in one drop of Canada balsam; the cover slip must be lowered slowly and carefully or bubbles of air will be trapped in the preparation;
(7) label slide: name of insect, part, method of preparation, date and your name.

EXPERIMENTS ON FEEDING

The mandibles, maxillae, labium and thoracic limbs are serially homologous. Draw them all on the same scale. Place the asymmetrical mandibles together and note the relation between their cutting and grinding surfaces. Make careful observations of a locust feeding; note the position of the legs and the part played by each of the mouthparts. Consider the structure of the mouthparts more carefully now that you have observed their function.

Experiment 9. Feeding Preferences

Prepare 0·008M, 0·15M and 0·30M sucrose solutions. Soak pieces of filter paper in each solution after marking each paper with a pencil to show molarity of sucrose solution. Also, as a control, soak some papers in distilled water. Dry the papers in an oven. Starve sixteen locusts overnight to empty their gut of food. Place these hungry locusts in separate containers. Cut the paper into shreds. Feed four locusts on 0·008M, four on 0·15M and four on 0·30M sucrose paper and four on the paper treated with distilled water. After 24 h collect, count, dry and then weigh the faecal pellets. The weight of faeces produced indicates the amount of food eaten (Dadd, 1960).

Experiment 10. Palpal Chemical Sense

Hold a hungry locust and observe its mouthparts under a low power binocular microscope. Hold a strip of filter paper (prepared as in Experiment 9) next to the labial and maxillary palps and, as you do so, observe the movements of the fore-limbs and mouthparts. Repeat several times and use a different locust for each paper.

Experiment 11. Feeding on Plants

Starve eight locusts overnight, to empty the gut, then place four in a container with the leaves of a graminaceous plant (grass) and four in a container with the leaves of a broad leaved plant (dandelion). After 24 h count the faecal pellets. Compare the results for *Locusta* and *Schistocerca,* if both species are available.

THE THORAX

The serial repetition of appendages (mandibles, maxillae and labium) is the only external indication of the segmental origin of at least part of the head. The three segments of the thorax are more apparent and there are three pairs of legs.

The first thoracic segment, or prothorax, has a large saddle-like tergum (pronotum). Find the two spiracles on each side of the thorax: one between the pro- and mesothorax (second thoracic segment), and the other between the meso- and meta-thorax (third thoracic segment). Look at the thorax from above to see the terga (dorsal plates), from below to see the sterna (ventral plates) and from the side (*Figure 12*) to see the pronotum of the prothorax and the pleura (lateral plates) of the meso- and metathoracic segments. Note that many of the exoskeletal plates are rigidly fused together.

Thoracic segments two and three, because the wings articulate with them, are referred to as the pterothorax (pteron = a wing). Compare the leathery tegmina or fore-wings (attached to the mesothorax) with the more delicate hind wings (attached to the metathorax). The membranous part of the wing is supported by long tubular veins and smaller cross veins. The four small axillary sclerites at the base of each wing can be seen if the thorax is viewed from above with the wings pinned out to the sides.

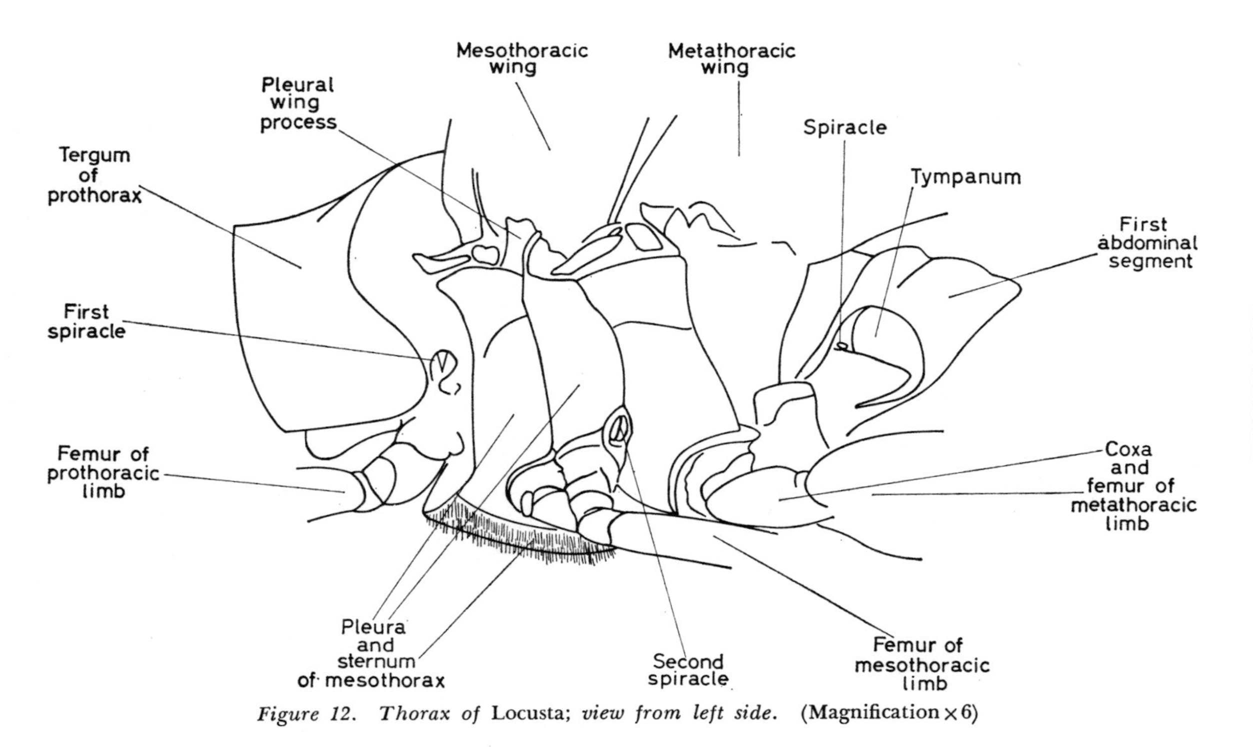

Figure 12. Thorax of Locusta; *view from left side.* (Magnification $\times 6$)

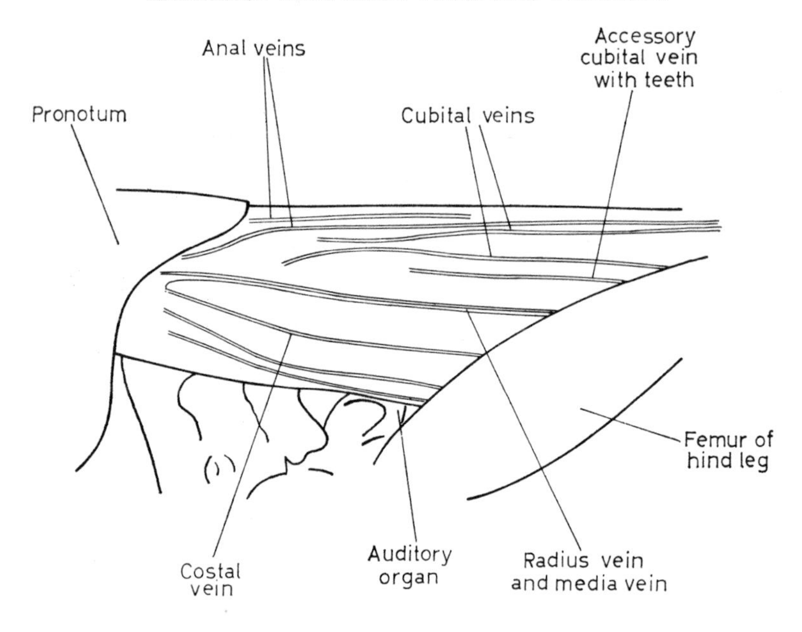

Figure 13. Fore-wing of male Locusta *in resting position.* (Magnification × 4)

Draw the tarsus and pretarsus of one leg to show the arrangement of the pads and claws (*Figure 14*). Can a locust climb up a smooth surface such as glass?

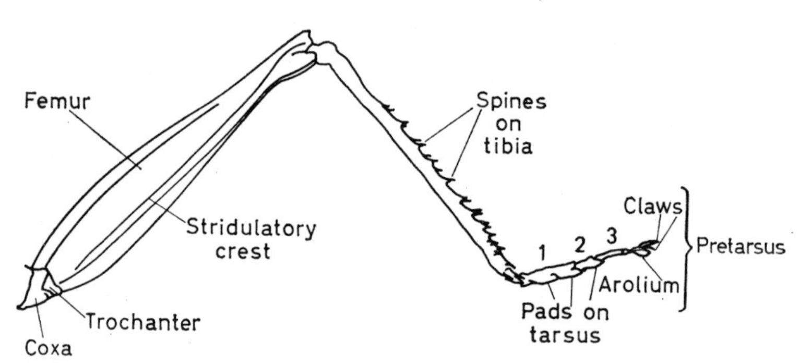

Figure 14. Inner aspect of right hind limb of a male Locusta. (Magnification × 2)

Attempt to relate the structure of the limbs to their movements. Note to what extent the parts of a limb can move in relation to each other. Try to see the movements of the limbs in walking, jumping, preening, feeding, stridulating, and in other behaviour. In stridulation (sound production—*see* Haskell, 1962) a ridge on the surface of the hind femora (*Figure 14*) is moved against a row of very fine teeth on the accessory cubital vein (*Figure 13*) of the fore-wing. This row of teeth is better developed in the male and only the male stridulates.

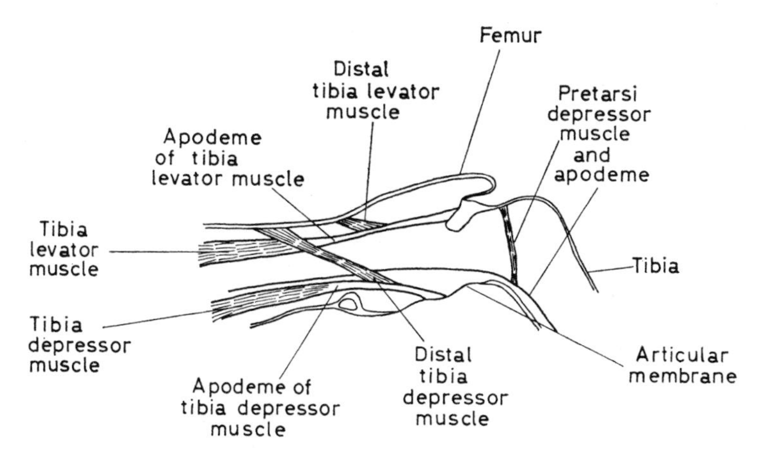

Figure 15. Joint between femur and tibia of right hind limb of Locusta, *cut longitudinally*

Cut the joint between the femur and tibia of a metathoracic limb in a vertical longitudinal bisection (*Figure 15*) and investigate the musculature. Pull on the muscles with your forceps and note their antagonistic effects. The tibia levator muscles provide the force of the jump. Contrast the external anatomy and the musculature of this limb with that of the pro- and mesothoracic limbs which take the shock at the end of a jump.

EXPERIMENTS ON BEHAVIOUR

The head and thorax, though structurally distinct, are closely related functionally. The thorax is a locomotory centre. In locomotion the head goes first and, here, important sense organs, structures used in feeding and the brain are found.

Experiment 12. Activity

Cloudsley-Thompson (1955) gives details of the designs of various types of aktograph apparatus. The locomotory movements of a locust in a small pivoted container may be recorded,

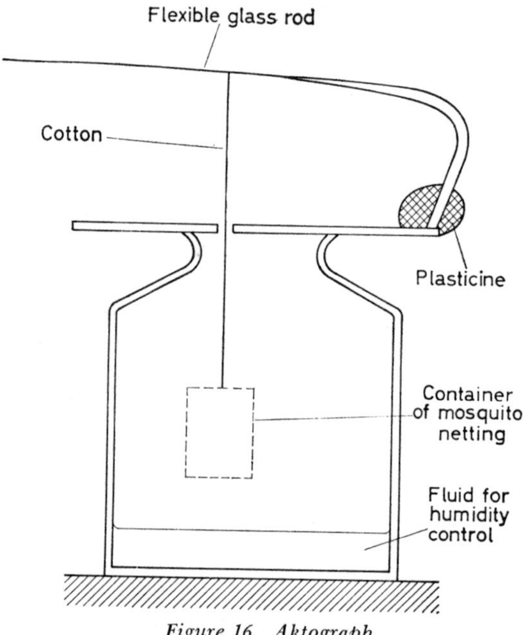

Figure 16. Aktograph

by way of a marker, on a kymograph drum or barograph chart. As far as possible, isolate the apparatus from vibrations and visual disturbance.

The activity of the hopper stages in a mosquito netting container (*Figure 16*) causes the marker to vibrate and mark a revolving smoked drum (for greater detail *see* Bursell, 1959).

An aktograph can be used to determine: (1) whether locusts are most active at 25°C, 30°C or 35°C (for more complex aktograph experiments *see* Edney, 1937 and Chapman, 1954); (2) whether hunger affects activity. Ensure that each locust is fed on fresh grass immediately before the experiment. A locust should be found more active at higher temperatures and as it becomes hungry.

Experiment 13. Tarsal Inhibition of Flight

Tie a piece of fuse wire around the thorax of an adult *Locusta* (between the first and second pairs of legs). Use the wire to lift

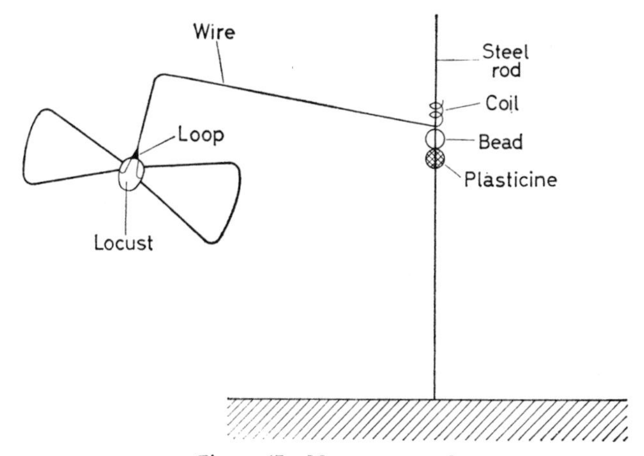

Figure 17. Merry-go-round

the locust; what happens? Replace the locust on the bench; what happens? In this experiment loss of contact with the ground affects flight.

Experiment 14. Locust Merry-go-round

See Figure 17; as the insect flies the coil rotates on the glass bead. Flight should be more sustained than in Experiment 13 (for greater detail *see* Sotovalta, 1955).

Experiment 15. The Waning of a Response

Hold a locust by its thorax; stroke the underside of the abdomen with a camel-hair brush and note the movement of the hind limbs. Repeat and note that, with repetition, the response wanes (Uvarov, 1928).

Experiment 16. Temperature Preference

Cover a sheet of glass with black card and use a 60W lamp to warm the glass from below. Confine some hoppers on the card; after a while they will have arranged themselves in a circle. Measure the temperature immediately above the bulb and where the locusts are resting. Repeat this experiment with a 100W lamp (Uvarov, 1928).

Experiment 17. Movement with Respect to Light

The observation cage should be long and narrow (*Figure 18*) and surrounded by a black cloth screen. Light entering at one

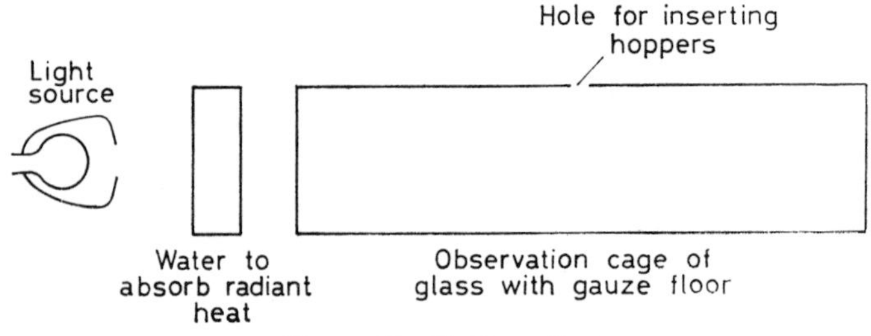

Figure 18. Light gradient

end of the cage through a hole in the screen is first passed through water which absorbs radiant heat. A very small observation hole is cut in the screen. Insert several locust hoppers through a hole in the centre of the cage roof and observe their behaviour. Record their position after 5 min. As a control leave the light switched off, insert the same number of hoppers and then, after 5 min, switch on the light. Record the position of the locusts. Compare the results (*see* Chapman, 1954).

Experiment 18. Two Light Experiment

If a dark room is available set up a bench as shown in *Figure 19*. Use one hopper at a time; place it at the release point and record its movements on graph paper. The experiment is best carried out at about 25°C so that the hopper is not too active.

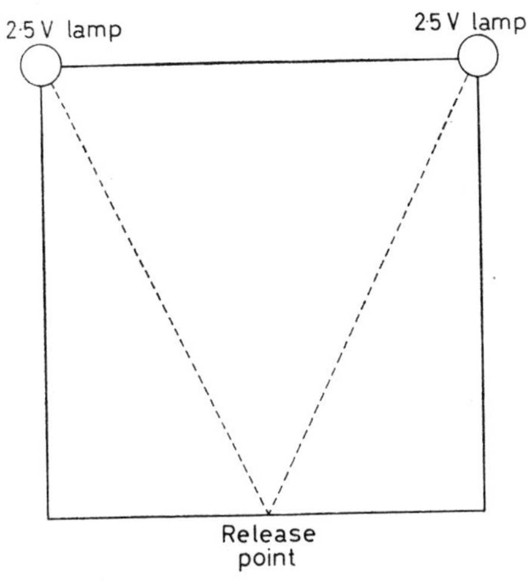

Figure 19. Two light experiment

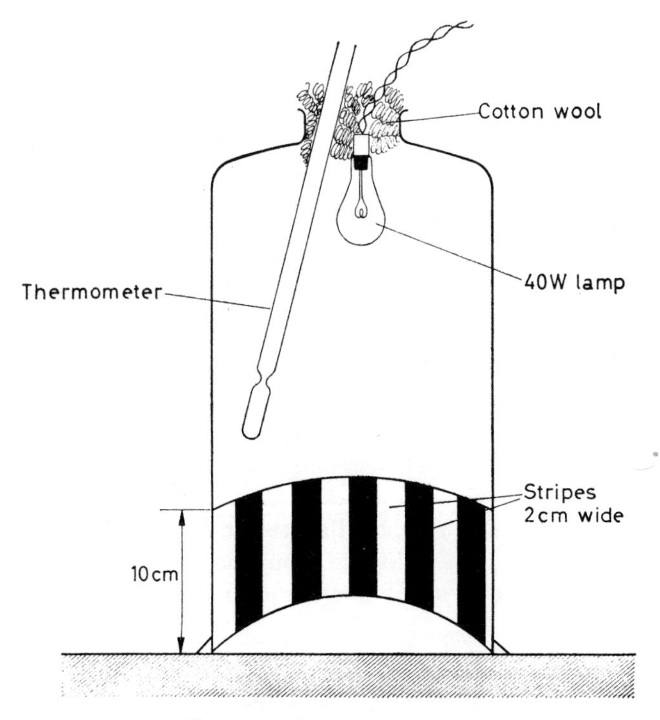

Figure 20. Optomotor response

Experiment 19. The Optomotor Response

The locust will move its body to compensate for the movement of stripes in its visual field. Enclose a locust hopper in a bell jar (*Figure 20*). Encircle the base of the bell jar with black and white striped paper. Switch on the light and record the temperature. Move the striped paper in either a clockwise or an anti-clockwise direction and observe the locust's behaviour. Below 25°C the movement of the head will be conspicuous. At about 29°C the locust will walk as the stripes move. If the stripes move in a clockwise direction, which way does the locust move?

Experiment 20. Marching Hoppers

Cut a hole (10 cm diameter) in the lid of a biscuit tin (23 × 23 × 23 cm). Also cut an observation hole (1 cm diameter) at one side of the lid. Cover the floor with white paper. Starve thirty mid-fourth or fifth stage *Locusta* hoppers overnight or for at least 2 h, then put them into the tin. These hoppers can be recognized as being in the middle of an instar if they are provided with fresh grass. Hoppers which have just moulted, or which are just about to moult, will not feed. Put a clear glass dish over the large hole in the lid and pour water into the dish to a depth of about 5 cm. Place a 40W lamp centrally above this. Use bench lamps outside the tin to raise the temperature inside to 35°C; then switch these off. Observe the hoppers. When they are marching several will walk together round and round the cage. They march best at 30–33°C (Ellis, 1951).

THE ABDOMEN

The abdomen comprises eleven segments; note the tympanal (auditory) organ in the first segment and the spiracles in segments 1–8. The exoskeletal plates of some segments overlap those of the next. Take a freshly killed locust, hold the thorax and pull on the tip of the abdomen. Note the extreme extensibility of the intersegmental membranes. Note the membranes between the terga and sterna.

Distinguish between the sexes (*See figure 8*) and observe the genitalia of a mating pair. Compare the cerci of the male with those of the female. The anus is below the epiproct and between the paraprocts.

EXPERIMENTS ON RESPIRATION AND EGG-LAYING

Experiment 21. Opening of Spiracles

The spiracles may be observed through a dissection microscope if the locust is fixed between pads of cotton wool in a small observation cell. The spiracle between the meso- and metathoracic pleura is large and can be seen as its valves move apart. Opening of the spiracles can be induced experimentally by increasing the carbon dioxide concentration of the atmosphere. This response is probably a result of the greater acidity of the tissues. Observe a spiracle as the carbon dioxide concentration increases (as a result of the locust's respiration) when a locust is in a small air-tight tube.

Experiment 22. Respiratory Movements

Observe a resting locust and note the movements of the abdomen which affect ventilation of the tracheal system. Count the number of movements in 1 min; is this affected by temperature?

Experiment 23. Egg-laying

Watch the locust cage until you see a female which is continually raising the tip of its abdomen and then pressing it against the floor. This means it is about to lay. Put some moist sand between two vertical sheets of glass (about 15 mm apart) so that the surface of the sand forms part of the cage floor. Observe the use of the ovipositor valves in digging, the extension of the abdomen, the laying of eggs (oviposition) and the flow of frothy secretion (from the accessory glands).

Experiment 24. Egg-laying and Soil Moisture

Heat some sand to drive off all the water. When the sand is cool divide it amongst several 1 lb. jam jars. Leave one jar unwatered and mix varying amounts of water with the sand in the other jars. Place the jars in the locust cage so that the sand is flush with the false floor. Examine the jars on the next day. You may find that the locusts have dug holes in some jars without laying and that they have laid in others. Locusta will not lay in either dry or very wet sand.

Experiment 25. Egg-laying and Soil Salinity

Mix one part 0·3M NaCl solution with five parts clean dry sand. Also mix one part distilled water with five parts clean dry sand. In which do the locusts lay eggs? Do they avoid the other altogether?

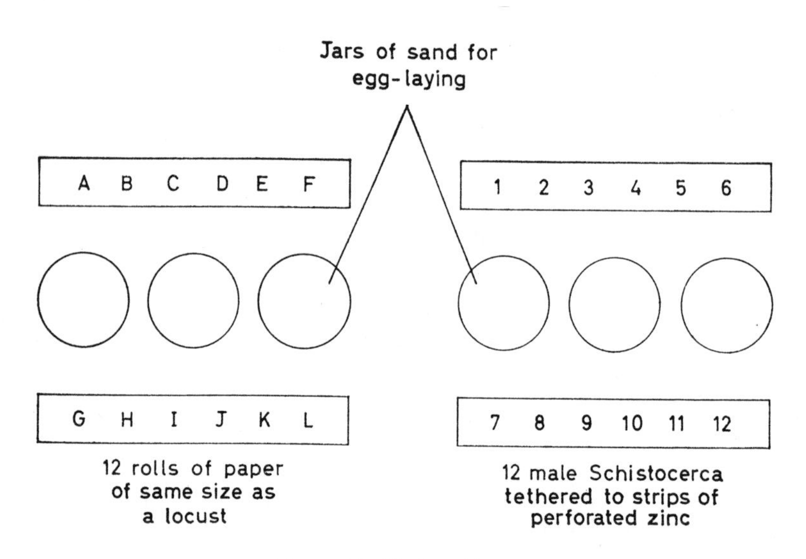

Figure 21. Tethered locusts

Experiment 26. Gregarious Behaviour of Egg-laying Females

Schistocerca must be used in this experiment. Place the jars for egg-laying as shown in *Figure 21.* Tether twelve locusts on two strips of perforated zinc and place them next to one group of jars. To tether each locust wrap a piece of fuse wire around the thorax, twist the wire once, pass the ends through the perforated zinc and twist again. Introduce females, which are just about to lay (see Experiment 23), and leave them overnight. In which jars are eggs laid? (*See* Norris, 1963.) The twelve rolls of paper should provide obstacles of about the same size as a locust.

Experiment 27. Digging without egg-laying

Fill a jar with dry sand to within about 5 mm of the top, and then use dry blue sand (which has been dyed with methylene blue) for about the last 5 mm. When a locust digs, the blue surface sand falls into the hole as the abdomen is withdrawn if eggs are not laid. When the blue surface sand is carefully removed later, the number of holes that have been dug since the experiment was set up can be counted. This demonstrates that in egg-laying behaviour, the earlier stages in the sequence are not necessarily followed by the later stages. The behaviour pattern may stop at any point if, for example, conditions are unsuitable.

Internal Anatomy

ARTHROPOD CHARACTERISTICS

Figure 22 is a drawing of a sagittal section of a locust; note the relations of parts inside the body. The following features are characteristic of all arthropods:

(1) The division of the alimentary canal into fore-gut (stomodaeum), mid-gut (mesenteron) and hind-gut (proctodaeum). The stomodaeum and proctodaeum are of ectodermal origin: note the cuticular teeth in the stomodaeum.
(2) The nervous system with dorsal supra-oesophageal ganglia (brain), ventral sub-oesophageal ganglia and a ganglionated ventral nerve cord (ganglion = swelling).
(3) The heart, dorsal in position and opening into the body cavity (haemocoel).
(4) The muscles attached to cuticular inpushings (apodemes).

INSECT CHARACTERISTICS

(1) The opening of the salivary duct posterior to the mouth (*Figure 23*).
(2) The excretory organs are Malpighian tubes.
(3) The genital ducts open posteriorly near the anus.
(4) The tracheae. These are hollow tubes which carry air into the tissues by diffusion and/or active ventilation. The blood (haemolymph) is relatively unimportant, therefore, in oxygen transport.

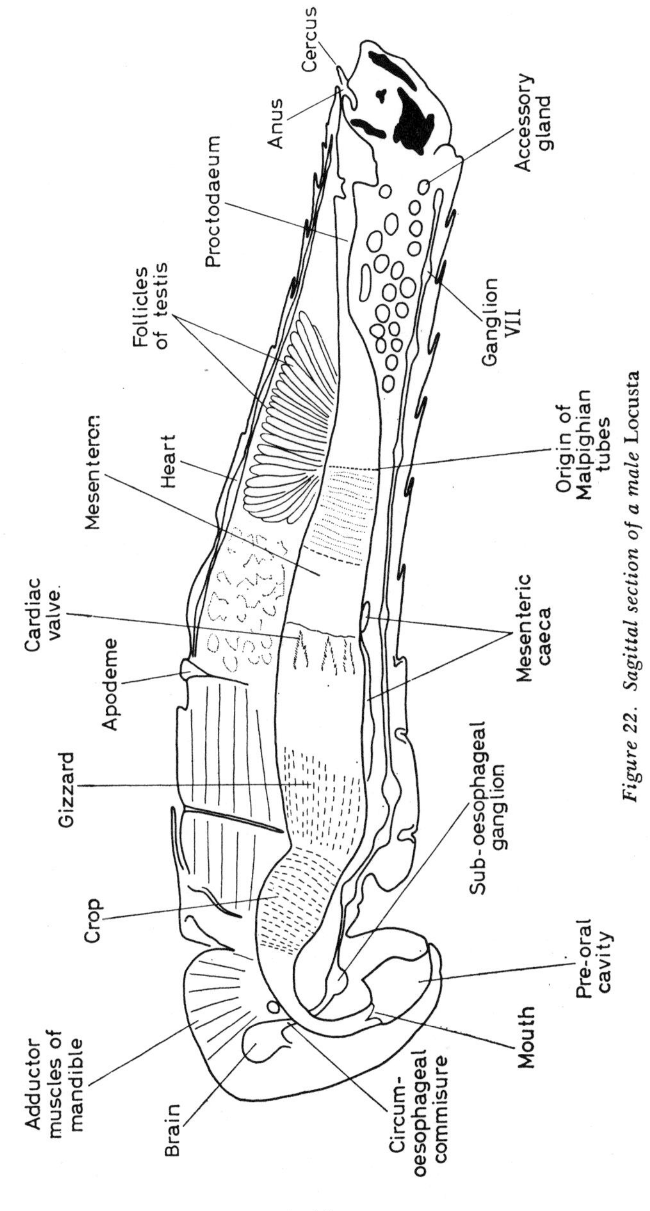

Figure 22. Sagittal section of a male Locusta

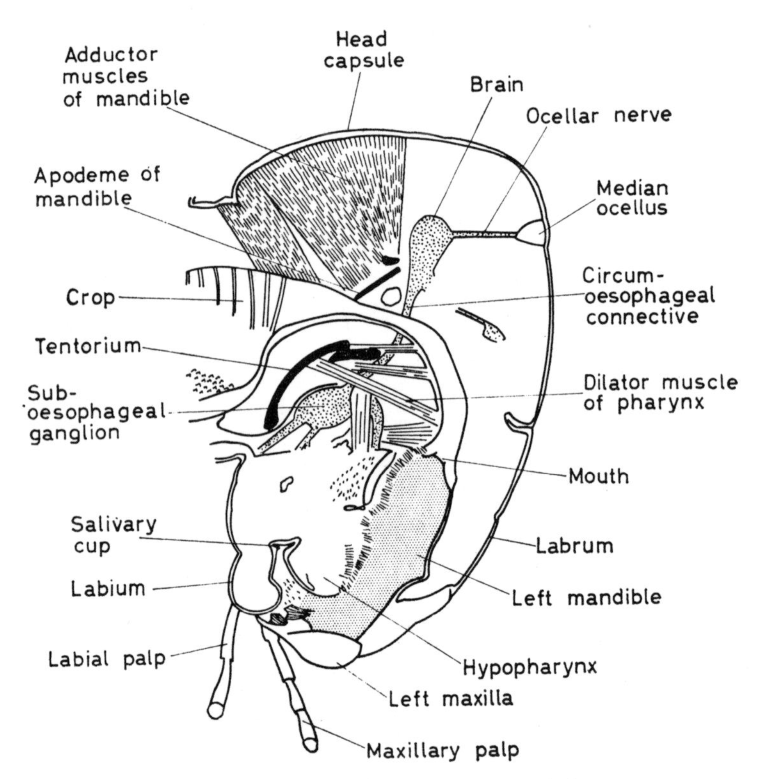

Figure 23. Sagittal section of the head of a male Locusta.
(Magnification ×9)

KILLING LOCUSTS

The substances used for killing insects (50: 50 mixture of chloroform and ether) are harmful to man and should be used carefully. Insects may be preserved for later dissection in a mixture of one part glycerine to one part 70 per cent alcohol.

GENERAL DISSECTION

Remove the wings and make a dorso-lateral cut along the right side of the abdomen. Continue the incision anteriorly, above the typanum and along the thorax. Place the ventral surface on the wax of a dissecting tray and pin through the femora of the legs. Pull on the abdomen to expose the intersegmental membranes and then pin through the external genitalia. Cover with water. Refer to *Figure 24*.

Use fine scissors to cut the dorso-ventral thoracic muscles of the right side below your incision through the cuticle. Turn the dorsal surface of the abdomen and thorax over to pin it down on the insect's left side. You are now looking at the heart from below and at the alimentary canal from above.

Alimentary Canal and Nitrogenous Excretory System

The Malpighian tubes are very fine threads extending both anteriorly and posteriorly around the gut; they originate at the boundary between the mesenteron and the proctodaeum. Follow the alimentary canal forwards and note the mesenteric caeca at the junction between the mesenteron and stomodaeum. The larger anterior caeca are applied to the gizzard. Anterior to this, note the crop and oesophagus. Pin the alimentary canal to one side and note the numerous white lobes of the salivary glands (only the left salivary gland is shown in *Figure 24*).

Transport Systems

The heart runs the length of the abdomen; it is a mid-dorsal tube with segmental dilations. Blood (haemolymph) enters the heart, from the haemocoel (body cavity), through an opening (ostium) on either side of each dilation. These ostia cannot be demonstrated easily. The alary muscles and dorsal diaphragm

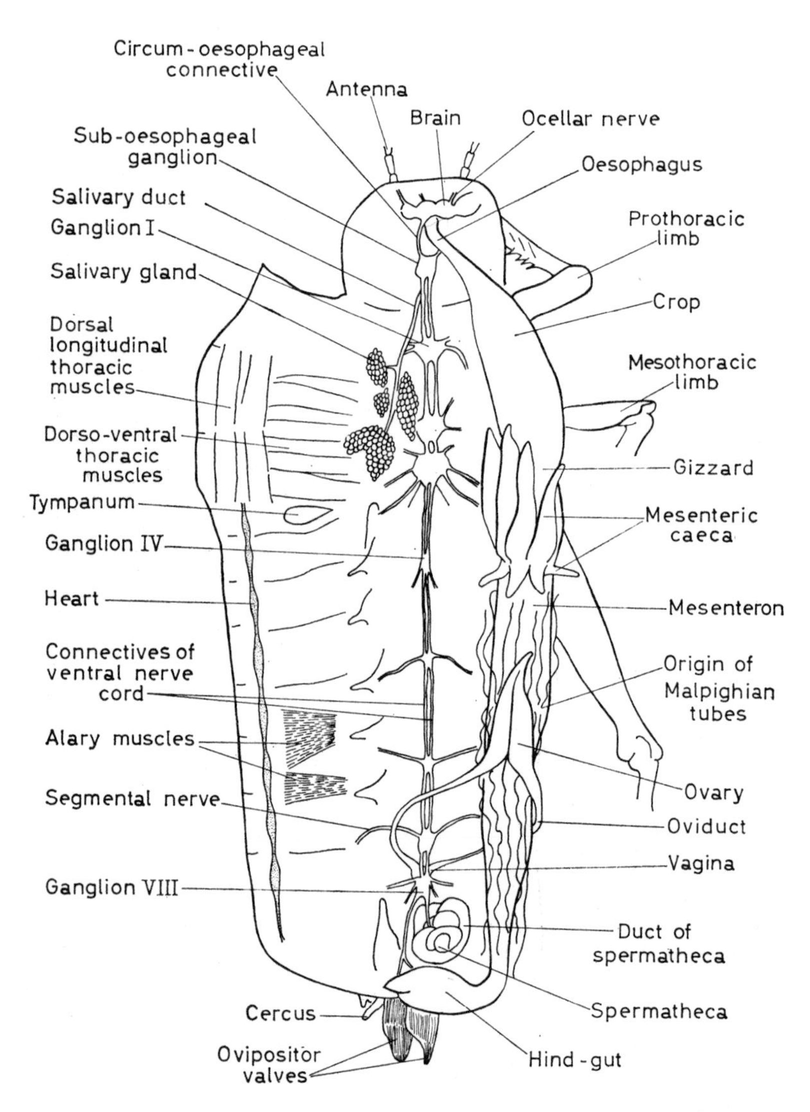

Figure 24. General dissection of a female Locusta

are just vental to the heart. The dorsal diaphragm, therefore, separates the perivisceral cavity from the pericardial cavity.

The part of the dorsal vessel which lies in the thorax and head is called the aorta; it is very delicate and is not usually seen in elementary dissections. Blood, carried anteriorly in the heart and aorta, is returned to the haemocoel in the head through an anterior funnel-shaped opening of the aorta.

Dorsal longitudinal tracheae lie on either side of the heart and are joined by transverse connections, dorsal to the heart. Note the much branched tracheal system with dilations, the air sacs. The air sacs are conspicuous soon after ecdysis when they occupy space in the haemocoel which is later occupied by other internal organs (especially the gonads).

As you proceed with the dissection, you will find it necessary to remove much of the tracheal system and the diffuse, pale yellow/white fat body.

Reproductive System

Mature female—The ovaries are dorsal to the gut. They are closely associated and each contains about 45 ovarioles. Each ovariole opens into either the right or left oviduct. It contains developing eggs and the most mature egg is nearest to the oviduct. The two accessory glands are anterior extensions of the oviducts. Posteriorly, the oviducts join below the ventral nerve cord. Pin the hind-gut to one side so that you can see the oviducts converging below the nerve cord. The vagina is a single tube passing below the nerve cord to the genital chamber. The coiled tube, between the large apophyses of the ovipositor valves, is the spermathecal duct leading to the spermatheca (seminal receptacle). This duct passes below the nerve cord and opens into the genital chamber just dorsal to the vaginal opening. Examine the ventral surface of the ovaries and find the tracheae associated with them.

Mature male—The paired testes are closely associated and appear as a single yellow structure dorsal to the gut in the mid-line. Examine the testes and note the large number of tubular follicles. Each follicle opens, by way of a short slender vas efferens, into either the right or left vas deferens. The vasa deferentia are ventral to the testes; trace them posteriorly to where they open into the ejaculatory duct. Just anterior to these openings, 30 tubular accessory glands open into the ejaculatory duct. Two of these tubes, the seminal vesicles, end in small dilations.

Nervous System

Make lateral incisions into the head capsule and carry them forward between the compound eyes. Then remove the dorsal part of the head capsule between these cuts. In doing so the mandibular adductor muscles will be cut near their insertion upon the head capsule (exoskeleton). Carefully remove the mandibular adductor muscles by pulling upon them, piece by piece, using fine forceps. Pin through the apodemes of the mandibles

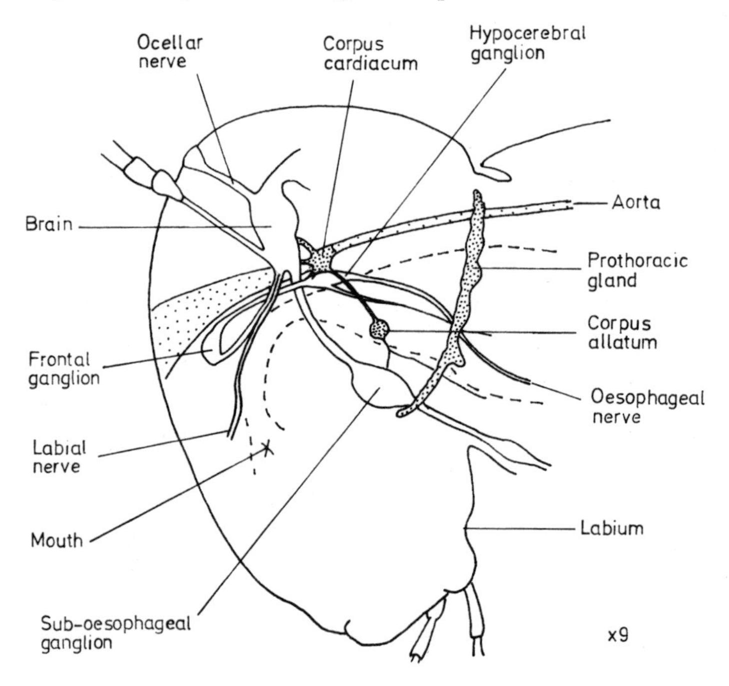

Figure 25. Locusta: *neuro-endocrine system in the head. Dissection from the side.*

so as to press the sides of the head capsule apart. Empty the water from your dissecting tray and cover the insect with 70 per cent alcohol; after some time the nerves become whiter. Leave the insect in alcohol overnight if possible.

Use fine forceps and a needle and carefully remove the ventral diaphragm which separates the perivisceral cavity from the perineural cavity. Trace the ventral nerve chain forwards; great care is needed in removing the fatty sheath which surrounds it. Note the abdominal and thoracic ganglia and connectives. Very carefully remove the tentorium (p. 42) by cutting through the tentorial arms on either side. Note the sub-oesophageal ganglia, circum-oesophageal connectives and supra-oesophageal ganglia. Trace the nerves associated with the ganglia. In the thorax note the first and second thoracic ganglia. The third ganglion in the chain is formed by the fusion of the third thoracic and abdominal ganglia 1, 2 and 3. The fourth, fifth, sixth and seventh ganglia in the chain are related to abdominal segments 4, 5, 6 and 7. The eighth ganglion is a fusion of the ganglia of abdominal segments 8, 9, 10 and 11.

Posterior to the supra-oesophageal ganglia and closely associated with the oesophagus, note the delicate oesophageal nerves of the sympathetic stomatogastric nervous system (*Figure 25*). Trace the outer oesophageal nerves posteriorly to the ingluvial ganglia on either side of the crop. Trace the oesophageal nerves anteriorly to the hypocerebral ganglion which is connected to the median frontal ganglion. Just dorsal to the hypocerebral ganglion find the corpora cardiaca (oesophageal ganglia) with fine nerves to the brain, to the hypocerebral ganglion, and to the conspicuous spherical bodies, closely associated with the oesophagus, the corpora allata. Hormones are secreted by both the corpora cardiaca and the corpora allata, and the latter are largest in older adults (14 days after ecdysis). The two *prothoracic* glands are also in the head; each is a long thin sheet of cells extending on either side of the oesophagus from near the neck membrane, dorsally, and towards the sub-oesophageal ganglion, ventrally. The prothoracic glands, which secrete juvenile hormone, are only conspicuous in the hopper stages and in *solitaria* adults. The stomatogastric nervous system and the hormone-secreting structures referred to above are best displayed in a dissection of the head from the side.

FURTHER STUDY OF THE ALIMENTARY CANAL

On completion of the dissection and of the related notes and drawings, open up the alimentary canal along its length. Note that the *pharynx* rises vertically from the *mouth,* and leads into the *oesophagus.* As the oesophagus passes posteriorly it gradually increases in diameter; its inner surface is covered with ridges and spines. The food is stored in the *crop* and (as a result of digestive juices passing forwards from the mesenteron) digestion commences. The longitudinal parallel ridges have minute conical spines in the anterior two thirds. The crop merges into the *proventriculus;* note the Y-shaped proventricular plates. In the *mesenteron* and *mesenteric caeca,* digestion continues and absorption takes place. In the *intestine* and *rectum* water is absorbed from the gut contents. Note the cuticular lining of the hind-gut, the longitudinal plates in the intestine and the rectal pads.

The waste products of nitrogen metabolism pass from the Malpighian tubes into the proctodaeum. Uric acid is the most usual nitrogenous excretory product of insects; it is non-toxic and can be excreted with little or no water. Water conservation is very important for the survival of land animals.

The anatomy and histology of the alimentary tract of *Locusta migratoria* is described in detail by Hodge (1939).

Experiment 28. Enzymes from the Alimentary Tract

Remove the gut from a freshly killed locust. Place the salivary glands, crop, gizzard, mid-gut and hind-gut in separate solid watch-glasses. Cut open and wash out the contents with water, and then grind up each part in 2 cm^3 water. Use these extracts to test for the presence of invertase, amylase, protease and lipase (Brown & Creedy, 1970).

Experiment 29. Movement of Food through the Alimentary Tract

The time taken for food to pass through the gut is related to the amount of food ingested each day. Starve locusts overnight and then place them on fresh grass in a container. Remove the locusts at different times after they have started feeding and place each one in a separate container so that a record can be made of the times when faeces are produced.

Experiment 30. Weight of Food Consumed

Divide some fresh grass equally between two similar containers. Place some locusts in one container for 24 hours. Divide the difference between the dry-weights of the grass from the two containers by the number of locusts used, to obtain an estimate of the dry weight of food consumed by a locust in one day.

DEMONSTRATION TECHNIQUES

(1) *Heart*

The *heart* is most clearly seen if the living locust is injected with an aqueous suspension of ammonia carmine 1 h before dissection. The dye is taken up by the nephrocytes on the surface of the heart (Albrecht, 1953). To prepare ammonia carmine mix ordinary carmine with 0·88 ammonia solution, and leave exposed to the air until partial putrefaction has taken place. Dry the product, then make up to 0·5 per cent solution with distilled water (Gurr, 1960).

(2) *Heart Beat*

Hold a newly fledged locust with its wings apart and observe, through the cuticle, the contraction of the heart. If necessary, inject with ammonia carmine to make the heart more conspicuous.

Experiment 31. Effect of temperature on heart beat

Animal hearts are either myogenic (heart beat not dependent upon innervation) or neurogenic (with nerve supply which is essential to the continued heart beat). To study, for example, the effect of temperature upon the rate of heart beat, remove the abdomen from a lightly anaesthetised locust. Cut long both sides of the abdomen to remove the roof of the abdomen in one piece. Wet some cotton wool with saline in a petri dish. Place the preparation on this pad. Examine the heart under a low power binocular microscope or use a hand lens. If kept moist with saline, the heart will continue to beat for a long time.

TABLE I Locust saline (Hoyle, 1953).

Molar	*Salt*	*Molar solution* ml	g *in locust saline*
58·5	NaCl	130	7·605
74·5	KCl	10	0·745
203·5	$MgCl_2 . 6H_2O$	2	0·407
219·0	$CaCl_2 . 6H_2O$	2	0·438
84·0	$NaHCO_3$	4	0·336
156·0	$NaH_2PO_42H_2O$	6	0·936

After mixing the solutions (except the $NaHCO_3$) add distilled water to make up to 1 l. Add the $NaHCO_3$ when the cylinder is nearly full, otherwise the Ca^{++} and Mg^{++} precipitate.

(3) *Tracheal System*

Put a freshly killed locust into a saturated solution of Sudan black in a mixture of equal parts of olive oil and kerosene (Pantin, 1960). The fluid displaces the air in the *tracheae.*

(4) *Organisms Living in Locusts*

Remove the alimentary tract from a freshly killed locust. Place the crop, mid-gut and hind-gut into separate watch glasses. Add two drops of saline to each and examine the gut contents under a microscope for *Protozoa.*

Nematodes (*Mermis* spp.) may also be present in the haemocoel.

(5) *Malpighian Tubes*

Remove the gut from a freshly decapitated locust and immerse it in a dilute solution of methylene blue in locust saline. The dye is rapidly taken up by the cells of the Malpighian tubes. Movements of the Malpighian tubes can also be observed.

Experiment 32. Excretion of Methylene Blue

Inject 0.5 cm^3 of a very dilute solution of methylene blue in locust saline into the haemocoele of a lightly anaesthetised locust. Faeces produced later will be blue (and the Malpighian tubes will be blue) until the methylene blue has been removed from the haemocoele. This demonstration works best if the insect is fed on filter paper (*see* Experiment 9) before and during the period of the investigation.

(6) *Nerve-muscle Preparations*

The frog sciatic nerve—gastrocnemius muscle preparation has long been used for class demonstrations but the locust is cheaper, is always available from biological suppliers, and its use may contribute to the conservation of natural populations of frogs.

Experiment 33. Electrical Stimulation of Nerves to the tibia-levator muscle

(a) Lightly etherise a locust. With scissors, remove the head, abdomen, wings, and the first two pairs of legs. Pull the remains of the gut from the thorax.

(b) Pin the thorax upside down on a cork sheet (*Figure 26*).

(c) To nullify the effect of contraction of the tibia-depressor muscle, cut through the apodeme (*Figure 15*) by making a shallow incision through the articular membrane.

(d) Fasten the legs by crossed pins, so that the tibia of one leg is free to move. Connect the end of this leg to a recording lever, by a cotton thread, so that the tibia is horizontal and contraction of the tibia-levator muscle can be recorded on a revolving drum.

(e) Use a sharp fine pointed scalpel to remove a rectangular piece of the sternal plate, to expose the metathoracic ganglion.

(f) Fix the leads from the stimulator to fine entomological pins. Pierce the ganglion with one of these pins, so that it is supported by the cuticle of the thorax. Insert the other pin into the femur. Keep the preparation moist with a drop of locust saline.

(g) Set the voltage control to 1·0 V and apply a single shock. Increase the voltage, if necessary, to give a maximum response. Set the recording drum speed to about 2·5 mm/sec, and arrange the marker so that it marks the drum.

Apply one shock per second. Slowly reduce the voltage, giving about 3 shocks at each voltage. When the leg no longer responds, increase the voltage again, step by step, and then decrease it again.

What is the effect on the response, of stimulating the preparation with different voltages in this way?

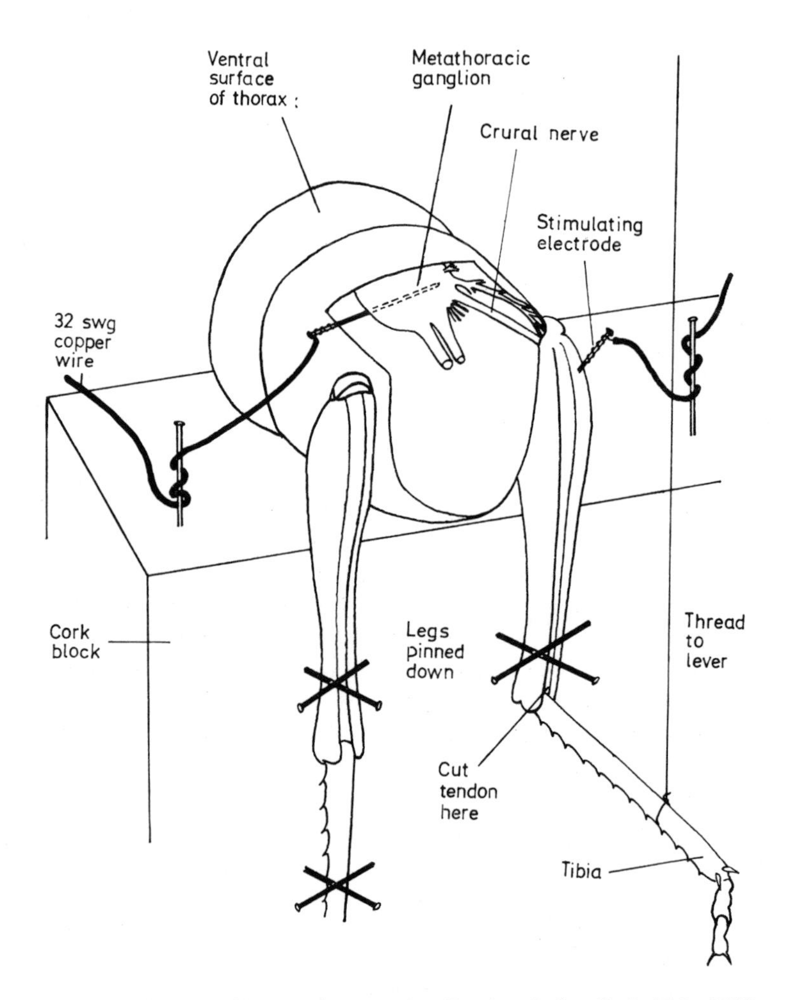

Figure 26. Diagram: Nerve-muscle preparation. Based on Barker, J. A. (Ed.) (1970), Control and Coordination in Organisms, *Harmondsworth; Penguin Books; and on Clark, R. B. (1966),* A Practical Course in Experimental Zoology, London; *John Wiley.*

(h) Set the voltage control to 1·0 V, if this gives a maximum response. Record the effect of 10 shocks at a frequency of one per second; then 10 at two per second; 10 at five per second; and 10 at ten per second.

What is the effect on the response, of different pulse rates?

(i) In this preparation, all the motor neurones to the leg are stimulated. What effect does it have on the record obtained, if the apodeme of the tibia depressor muscle is not cut in making this preparation?

How do you explain your observations in the last three stages [(g) to (i)] of this experiment?

(7) *Demonstration of Natural Colours in Museum Specimens*

When locusts are killed and set out in museum demonstrations they quickly lose their natural colours. To prevent this:

(a) Kill the locust by dropping it into alcohol.

(b) Make a lateral incision between the terga and sterna of the abdomen. Cut away the gut at its posterior end and pull at the anterior end to remove as much of the gut as possible. Remove the fat body and tracheae from the abdomen. Clean out any remaining tissues with cotton wool.

(c) Place a roll of dry cotton wool into the abdomen to preserve its shape.

(d) Dry the insect slowly for two to three weeks.

This method is recommended by Mr. Bill Page of The Centre for Overseas Pest Research, London.

Preparation of Permanent Mounts

INTRODUCTION

The stages in the preparation of permanent mounts of the mouthparts have already been given (p. 27). These stages have names which require explanation:

(1) Fixation is the killing of the tissue with the minimum of distortion.

(2) Dehydration: Living organisms contain large amounts of water which must be removed in permanent preparations. It is replaced by passing through increasing concentrations of alcohol. The water is finally removed in absolute alcohol; two changes are essential.

(3) Clearing: The clearing agent replaces the alcohol and increases the transparency of the tissue. Xylene or clove oil are frequently used. Two changes are necessary: if any water remains in the tissue the xylene becomes milky.

(4) Mounting: Clean slides and cover slips in acid alcohol before use. Mounting in Canada balsam facilitates microscopic examination and preserves the tissue. The Canada balsam is dissolved in xylene and it hardens as the xylene evaporates. The cuticle of the mouthparts has a colour which facilitates microscopic examination, but it is also possible to make permanent mounts of more or less transparent internal parts.

(5) Staining, either before or in the course of dehydration, renders transparent parts visible under the microscope. Also biological stains do not usually stain all parts to the same extent.

(6) Differentiation: A stain may be accepted particularly, for example, by the nuclei of a tissue. This is true of borax carmine and of haematoxylin. If a tissue is left in one of these until the nuclei are well stained but the cytoplasm is not deeply stained (progressive staining) the nuclei can be distinguished

from the cytoplasm. Alternatively, the tissue as a whole may be overstained and then the stain removed (regressive staining) from the cytoplasm. Both of these treatments result in differentiation. Differentiation may also be accomplished by using a second stain (a counter stain), of different colour, for the cytoplasm of the cells.

For greater detail on microscopical technique *see* Pantin (1960).

BASIC TECHNIQUES

Permanent Preparations Stained with Borax Carmine

Borax carmine stains nuclei red, whereas the cytoplasm remains pink or colourless. Kill a locust (do not dissect under water as this will cause the tissues to swell); remove a piece of trachea, an ovariole or a tubular filament from the testis, a Malpighian tube, a salivary gland, part of the wall of the crop, a very small piece of muscle (separate into fine fibres with a needle) and part of the ventral nerve chain.

(1) Fix in 70 per cent alcohol (10 min) in a watch glass.
(2) Place in 50 per cent alcohol (1 min). Always use a needle or forceps when transferring material, not a camel-hair brush.
(3) Stain in borax carmine (10 min).
(4) Wash in 50 per cent alcohol.
(5) Transfer to 70 per cent alcohol (1 min).
(6) Remove excess stain (differentiation) in acid alcohol (four drops concentrated HCl in 100 cm^3 70 per cent alcohol). The longer the material is in acid alcohol the more stain is removed (observe this under a microscope but the stage of the microscope must be kept dry).
(7) Transfer to 70 per cent alcohol, to 90 per cent, to 95 per cent, leaving for 1 min in each.
(8) To complete dehydration, two changes in absolute alcohol are essential. Leave for 5 min or longer in each and keep the watch glass covered.
(9) Clear in xylene; two changes of 5 min each are better than one.
(10) Mount separately in Canada balsam.

You will find that different material requires somewhat different treatment and the times indicated in parentheses above are intended only as a guide.

Permanent Preparations Stained with Ehrlich's Haematoxylin and Eosin

Haematoxylin stains nuclei blue and eosin stains cytoplasm pink. Kill a locust (do not dissect under water), remove either a small piece of muscle from the thorax or part of the ventral nerve cord.

(1) Fix in 70 per cent alcohol. Use as small a piece of muscle as possible and squash it to separate the fibres. Leave for 10 min.
(2) Stain in Ehrlich's haematoxylin (10 min); overstain.
(3) Differentiate in acid alcohol (until only nuclei are coloured); regressive staining.
(4) Wash in 70 per cent alcohol.
(5) Place in ammoniated alcohol until the nuclei are blue.
(6) Transfer to 70 per cent alcohol and then to 90 per cent alcohol; 1 min in each.
(7) Counterstain (progressive staining) in alcoholic eosin (1–2 min).
(8) Dehydrate: 90 per cent alcohol; 95 per cent alcohol; two changes in absolute.
(9) Clear in xylene: two changes (5 min each).
(10) Mount in Canada balsam.

Examination of Preparations

Trachea: note the nuclei of the squamous epithelium and the spiral thickening of the cuticle.

Ovariole and testis: the most mature sex cells are nearest to the genital ducts.

Malpighian tube: note the epithelial cells of the tube and their conspicuous nuclei.

Salivary gland: note glandular portions, the epithelial cells of the ducts and their spiral lining.

Wall of crop: note the cuticular teeth.

Muscle: insect muscle is striated.

Nerve chain: note ganglia, segmental nerves and connectives.

PREPARATION OF SMEARS AND SQUASHES

Wash slides in acid alcohol to remove grease.

Haemolymph

Prick the membrane at the base of the hind-limb of a living locust. Touch the drop of fluid which exudes on to a dry microscope slide. When this has become almost dry, fix by flooding with 70 per cent alcohol. Stain, dehydrate, clear and mount as outlined above (use Ehrlich's haematoxylin and eosin).

Spermatogenesis

Dissect out the testes from a freshly killed late 5th stage hopper (or from a recently fledged adult). Do not dissect under water. Examine the follicles in saline and clean off all the fat body and tracheae.

Method 1—Acetic Orcein (or Proprionic Orcein)

Proprionic orcein is easier to make up (1 per cent soln., G. T. Gurr's special synthetic orcein for chromosomes, in cold proprionic acid) and is more stable than acetic orcein. Transfer the follicles to 1 per cent acetic or proprionic orcein in a watch glass. Keep them covered; overstaining is not possible.

Either (a) *Examination of whole follicle*

Place two or three follicles on a microscope slide in a drop of fresh stain. Cover with a cover slip and then warm the preparation over a low flame until the slide is just too hot to bear on the back of your hand. Allow the slide to cool and then repeat the warming two or three times. Examine under the low power of a microscope and note the general architecture of the follicle and the sequence of stages in spermatogenesis. Apply gentle pressure to the cover glass so as to just burst the follicle whilst keeping the cells in approximately their correct sequence. Examine again and note that meiotically dividing cells are in localised groups. Note also the stages in mitosis in the production of primary spermatocytes.

Or (b) *Squash Preparation* (*Figure 27*)

Place two or three whole follicles on a microscope slide in a drop of stain. Cover with a cover glass. Press with the left thumb on top of the right thumb, with several layers of blotting paper between

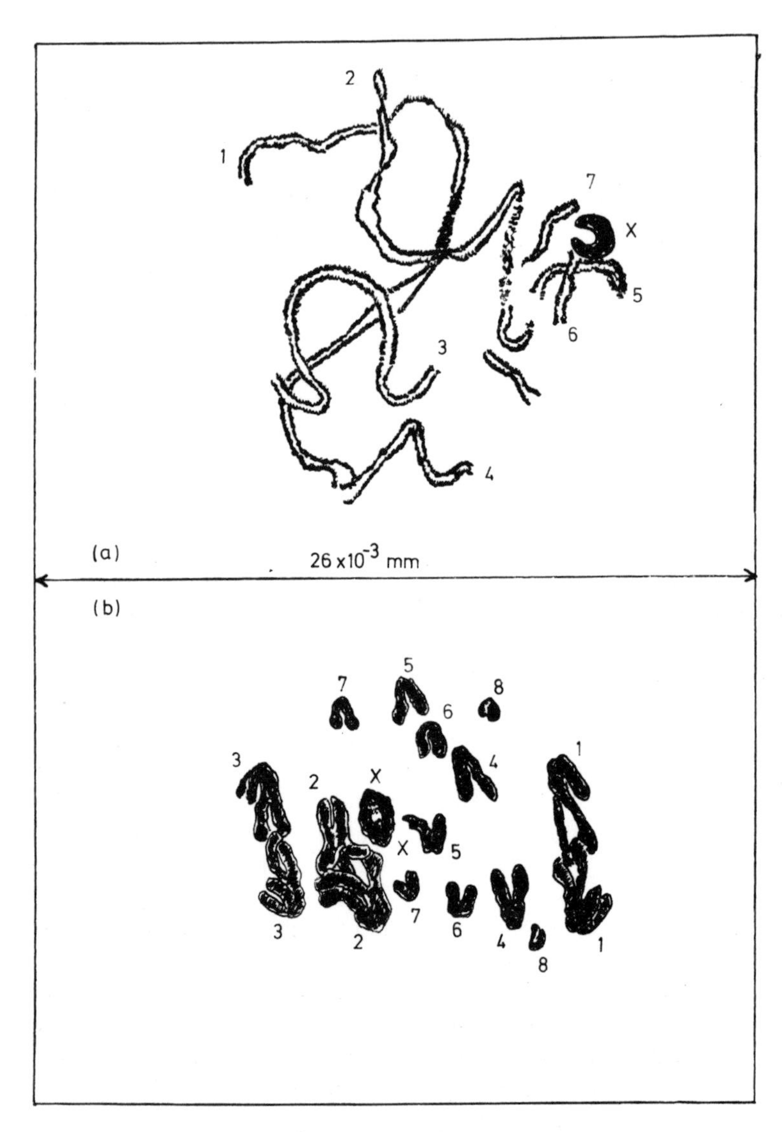

Figure 27. Stages in meiosis from the testis of a European grasshopper Chorthippus brunneus. *In* Chorthippus, *males have eight pairs of autosomal chromosomes and a single 'X' chromosome* (2n = *17*), *and females have two 'X' chromosomes* (2n = *18*). (*a*) *Pairing of homologous chromosomes in prophase of first division* (pachytene). *Pairs of chromosomes numbered, also 'X' chromosome.* (*b*) *Early anaphase of first division; members of each bivalent separated.*

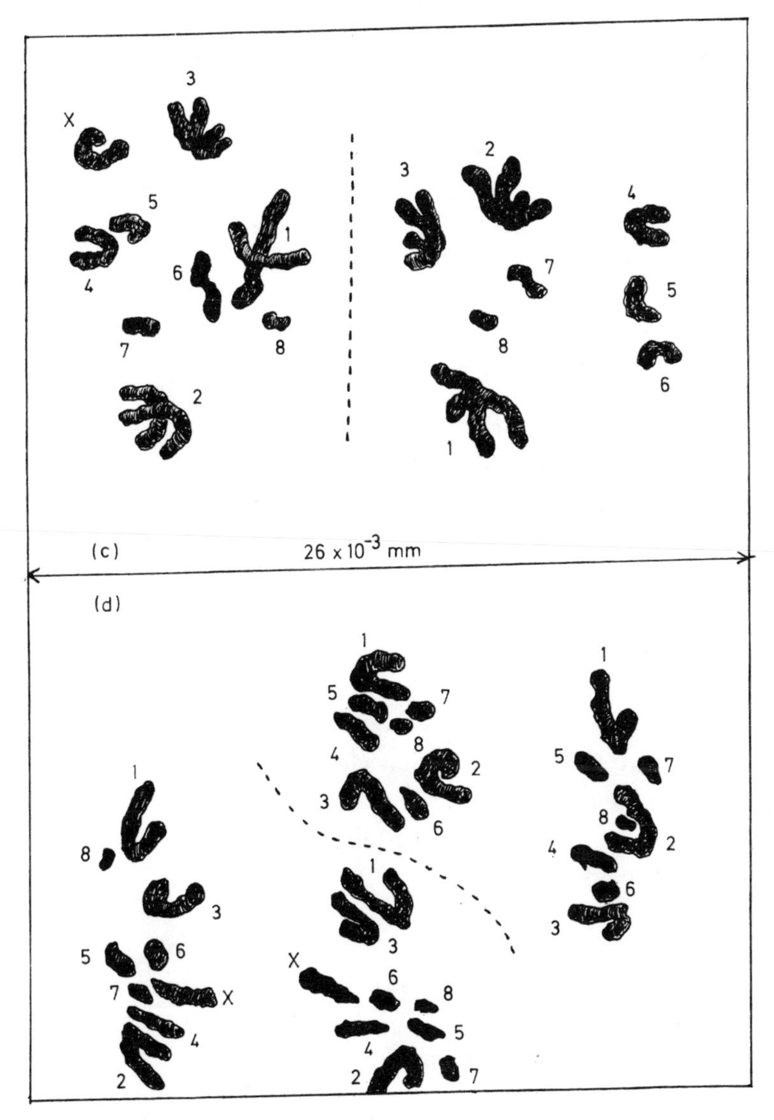

(c) Prophase of second division. Two cells indicated by a broken line; one cell with an 'X' chromosome and one without. (d) Anaphase of second division. Four sets of chromosomes.

Note that *in* Locusta (2n = *23 in the male) there may be up to four supernumerary chromosomes.*

Based on colour transparencies from Harris Biological Supplies (see *p. 63*).

the thumb and the cover glass. Avoid lateral movement and press down firmly. Either warm and cool as before but with up to ten repetitions, or support the preparation on glass rods over a bath of boiling water for 10 minutes. Seal with rubber solution (as used for cycle tyre repairs).

Method 2—The Feulgen reaction for D.N.A. (a histochemical technique for the study of nuclei in meiosis).

(1) Fix the smear in Navishin (100 parts 2 per cent chromic acid; 100 parts 20 per cent acetic acid; 80 parts 40 per cent formaldehyde and 20 parts distilled water); (5 min).
(2) Hydrolyse in N HCl at 60°C in a water bath (6 min).
(3) Stain in leuco-basic fuchsin (Schiff's reagent) (60 min).
(4) Rinse in water saturated with SO_2 from siphon (freshly prepared): three changes of 10 min each.
(5) Rinse in distilled water.
(6) Dehydrate, clear and mount.

Method 3—Crystal violet (stains chromosomes blue).
(1) Fix smear in an aqueous fixative (aqueous Bouin) (15 min).
(2) Stain in 0·5 per cent crystal violet (3–10 min).
(3) Rinse in distilled water.
(4) One part iodine in 99 parts of 1 per cent KI in 80 per cent alcohol (30–45 sec).
(5) Rinse in 90 per cent alcohol.
(6) Absolute alcohol (4–10 sec).
(7) Differentiate under microscope: clove oil (30 sec).
(8) Xylene: three changes of 10 min each.
(9) Mount in neutral balsam.

For greater detail on the handling of chromosomes *see* Darlington and La Cour (1969); for biological laboratory data *see* Hale (1966).

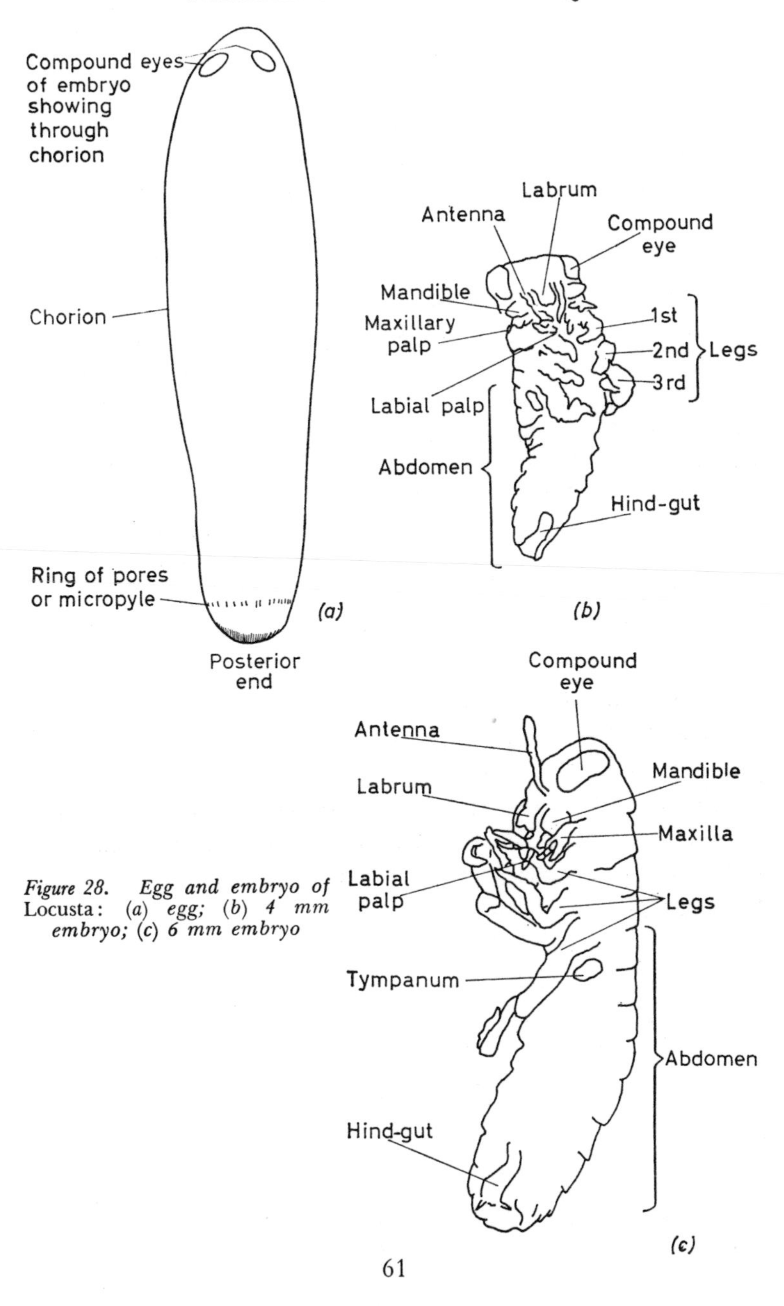

Figure 28. Egg and embryo of Locusta: *(a) egg; (b) 4 mm embryo; (c) 6 mm embryo*

EMBRYOS

Examination of Living Embryos

When the embryo is about 5 mm long the eyes are slightly pigmented and movements of the embryo can be seen through the chorion (egg shell). To remove the chorion from living grasshopper eggs (Slifer, 1945) immerse the eggs in 2 per cent sodium hypochlorite and then wash several times in fresh water.

Permanent Stained Preparations of Embryos (Figure 28)

Incubate the eggs at a constant temperature. The locust egg (*Figure 28*) has a ring of pores, the micropyle, at its posterior end and fertilization occurs as the egg leaves the ovary. Remove any sand from the egg and open it on a microscope slide. Prick with a needle near the anterior end and hold; press a seeker on to the posterior end and pull the egg away from the needle. This removes the anterior end. Now hold the needle on its side, next to the seeker, and push it forward to squeeze out the yolk. The embryo floats out with the yolk:

(1) Fix on the slide in 70 per cent alcohol (5 min).

(2) Remove excess yolk very carefully.

(3) Flood with 50 per cent alcohol (2 min).

(4) Flood with borax carmine (5–10 min).

(5) Dehydrate, clear and mount.

The times indicated are only approximate; the larger the embryo, the longer the time needed at each stage. Details of locust embryology are given by Roonval (1937), Hunter-Jones & Chapman (1964).

Appendix

VISUAL AIDS

Colour Transparencies

Ecdysis in Locusta. 12 slides illustrate the sequence of emergence of the imago from the fifth hopper stage, and also features of the anatomy of the imago. Obtainable from Harris Biological Supplies Ltd., Oldmixon, Somerset.

Meiosis in the testis of Chorthippus. 12 slides illustrating stages in meiosis. Obtainable from Harris Biological Supplies Ltd.

8 slides of the locust have also been produced by the Nuffield Foundation for the 'O' level Biology project.

16 mm Films

The Ruthless One. A Shell Film (20 min). Considers: body form; sex differences; mating; oviposition; life-history; numbers; food consumption; control.

The Rival World. A Shell Film (20 min). The dramatic presentation of the locust problem is only a part of this film but its impact is seen in relation to the vast field of economic entomology.

Shell films are available on free loan from: Petroleum Films Bureau, 4 Brook Street, London, W.1.

The Desert Locust. A World-Wide Picture (18 min). History of locust problem; external features, life-history and habits of *Schistocerca*; size of swarms and food consumption; methods of control and the need for international co-operation. On hire from: Central Film Library, Government Building, Bromyard Avenue, Acton, London, W.1.

Locusts at Hawthorndale. I.C.I. (6 min). Locusts reared in large numbers and used in testing insecticides. Close-up shots of adult locusts feeding and of the emergence of vermiform larvae from the sand. On free loan from I.C.I. Film Library, Imperial Chemical House, Millbank, London, S.W.1. Ask for sound version.

Le Monde Sonore des Sauterelles. Institut Français (45 min). Sound production, reactions to sounds, locus of response. Rental from Institut Français, 15 Queensberry Place, London, S.W.7.

Phase contrast microscopy: Spermatogenesis. Dr. Kurt Michel (27 min). Shows clearly the behaviour of chromosomes in spermatogenesis in a grasshopper. On hire from: British Film Institute, 81 Dean Street, London, W.1.

8 mm Film loops; Nuffield 'O' level biology project.

BIOLOGICAL SUPPLIES

Living Locusts

Supplies are obtainable, in Britain, from the following biological supply firms: GBI (Labs) Ltd., Heaton Street, Denton, Manchester. Gerrard Haig & Co. Ltd., Worthing Road, East Preston, Nr. Littlehampton, Sussex. Griffin Biological Laboratories, 113 Lavender Hill, Tonbridge, Kent. Harris Biological Supplies Ltd., Oldmixon, Weston-super-Mare, Somerset. Worldwide Butterflies Ltd., Over Compton, Sherbourne, Dorset. Larujon Locust Supplies, Welsh Mountain Zoo, Colwyn Bay, North Wales. The Butterfly Farm, Bilsington, Ashford, Kent.

Preserved Orthoptera

Supplies are obtainable in North America, as follows: *Schistocerca*, *Melanoplus*, *Romalea*, *Gryllus*, and *Gryllotalpa*. Suppliers: Carolina Biological Supply Co., Burlington, North Carolina 27215. General Biological Supply Co., 761–763 East 69th Place, Chicago 37, Illinois. Wards of California, P.O. Box 1749, Monterey, California 93940. Ward's Natural Science Establishment, 3000 Ridge Road East, Rochester 9, New York. Robert C. Wind, 827 Congress Avenue, Pacific Grove, California. Turtox Laboratories, 8200 South Hayne Avenue, Chicago, Illinois 60620.

Prepared Microscope Slides

Locust section through eye (Harris Z7-404); Locust T.S. thorax (Harris Z7-405; Griffin 8Z-150); Locust T.S. abdomen (Harris Z7-406; Griffin 8Z-150); Locust V.L.S. 4th. hopper stage (Harris Z7-407); Locust embryo stained acetic orcein for mitosis (Harris

Z7-414); Locust testis L.S. follicle for stages in spermatogenesis (Harris Z7-412); Locust testis squash for meiosis (Harris Z7-413; GBI 5550Z; Griffin 8Z-152); *Chorthippus* testis squash (Harris Z7-416); Grasshopper testis section (Gerrard GE880); Grasshopper sperm smear (Gerrard GE877).

Locust Cages

Harris B5026; egg tubes B5026/10; Griffin L05-740; egg tubes L05-752/005; also from Gerrard Haig, and from Aluminium Equipment Ltd., 21a Conewood Street, London N5 1BZ.

Rearing Cylinders

Suitable for hopper stages or for up to about 50 adults: GBI (Labs) 244D; Gerrard Haig AG2160; Griffin L04-110; Saruman Butterflies, 58 High Street, Tunbridge Wells, Kent.

Specialists in Entomological Equipment

Watkins & Doncaster, 110 Park View Road, Welling, Kent.

FURTHER READING

For further information on the classification, anatomy, physiology and natural history of insects, see:

Imms, A. D. (1947). *Insect Natural History*. London; Collins.

— (1961). *Outlines of Entomology* (revised by Richards and Davies). London; Methuen.

Wigglesworth, V. B. (1964). *The Life of Insects*. London; Weidenfeld & Nicolson.

References

Albrecht. F. O. (1953). *The Anatomy of the Migratory Locust.* London; Athlone Press.

Brown, G. D. and Creedy, J. (1970). *Experimental Biology Manual.* London; Heinemann.

Bursell, E. (1959). 'The Water Balance of Tsetse Flies'. *Trans. R. ent. Soc. Lond.*, **111,** 205.

Chapman, R. F. (1954). 'Responses of *Locusta migratoria migratorioides* (R. & F.) to Light in the Laboratory'. *Brit. J. Anim. Behav.*, **2,** 146.

Cloudsley-Thompson, J. L. (1955). 'The Design of Entomological Aktograph Apparatus'. *Entomologist,* **88,** 153.

Dadd, R. H. (1960). 'Observations on the Palatability and Utilisation of Food by Locusts, with Particular Reference to the Interpretation of Performances in Growth Trials using Synthetic Diets'. *Ent. exp. appl.*, **3,** 283.

Darlington, C. D. and La Cour, L. F. (1969). *The Handling of Chromosomes.* (5th edn). London; Allen & Unwin.

Dirsh, V. M. (1950). 'A Practical Table for the Determination of Sexes of Nymphs of *Locusta migratoria migratorioides* (R. & F.) (Orthoptera, Acrididae)'. *Proc. R. ent. Soc. Lond.*, **19** (B), 136.

— (1953). 'Morphometrical Studies on Phases of the Desert Locust'. *Anti-Locust Bull.*, **16.**

Edney, E. B. (1937). 'A Study of Spontaneous Locomotor Activity in *Locusta migratoria migratorioides* (R. &. F.) by the Actograph Method.' *Bull. ent. Res.*, **28,** 243.

Ellis, P. E. (1951). 'The Marching Behaviour of Hoppers of the African Migratory Locust in the Laboratory'. *Anti-Locust Bull.*, **7.**

Gurr, E. (1960). *Encyclopaedia of Microscopic Stains.* London; Hill.

Hale, L. J. (1966). *Biological Laboratory Data.* (2nd edn). London; Methuen.

Haskell, P. T. (1962). *Insect Sounds.* London; Witherby.

Highnam, K. C. (1962). 'Hormones and Swarming in Locusts'. *New Scientist,* No. 283, 86.

Hodge, C. (1939). 'The Anatomy and Histology of the Alimentary Canal of *Locusta migratoria* L.'. *J. Morph.*, **64,** 375.

Hoyle, G. (1953). 'Potassium Ions and Insect Nerve Muscle'. *J. exp. Biol.*, **30,** 121.

Hunter-Jones, P. (1966). *Rearing and Breeding Locusts in the Laboratory.* London; Anti-Locust Research Centre.

— (1964). 'Egg Development in the Desert Locust (*Schistocerca gregaria* Forsk.) in Relation to the Availability of Water'. *Proc. R. ent. Soc. Lond.*, (A), **39,** 25.

REFERENCES

Hunter-Jones, P. and Chapman, R. F. (1964). 'Egg Development in Locusts'. *School Sci. Rev., Lond.*, **95,** 658.

Jackson, C. H. N. (1933). 'On a Method of Marking Tsetse Flies'. *J. anim. Ecol.*, **2,** 289

Kalmus, H. (1948). *Simple Experiments with Insects*. London; Heinemann.

Norris, M. J. (1963). 'Laboratory Experiments on Gregarious Oviposition in the Desert Locust (*Schistocerca gregaria* Forsk.)'. *Anim. Behav.*, **11,** 408.

Oldroyd, H. (1970). *Collecting, Preserving and Studying Insects*. (2nd edn.) London; Hutchinson.

Pantin, C. F. A. (1960). *Notes on Microscopical Technique for Zoologists*. Cambridge University Press.

Roonval, M. L. (1937). 'Studies on the Embryology of the African Migratory Locust'. *Phil. Trans.* (B), **227,** 175.

Slifer, E. H. (1945). 'Removing the Shell from Living Grasshopper Eggs'. *Science,* **102,** 282.

Solomon, M. E. (1951). 'Control of Humidity with Potassium Hydroxide, Sulphuric Acid, or Other Solutions'. *Bull. ent. Res.*, **42,** 543.

Sotovalta, O. (1955). 'A Simple Method for Studying and Demonstrating the Energy Consumption in Flying Insects'. *Nature,* **175,** 543.

Uvarov, B. P. (1928). *Locusts and Grasshoppers*. London; Imperial Bureau of Entomology.

— (1966). *Grasshoppers and Locusts: A Handbook of General Acridology*. Vol. 1 Anatomy, physiology, development, phase polymorphism, introduction to taxonomy. London; Cambridge University Press.

— (1951). *Locust Research and Control* (1929–1950). Colonial Research Publication No. 10. London; H.M.S.O.

Wigglesworth, V. B. (1966). *Insect Physiology* (6th edn.) London; Methuen.

INDEX

Italicized page numbers indicate reference to a diagram

INDEX

Date Due